MEMBRANE PROCESSES IN INDUSTRY AND BIOMEDICINE

MEMBRANE PROCESSES IN INDUSTRY AND BIOMEDICINE

Proceedings of a Symposium held at the 160th National Meeting
of the American Chemical Society, under the sponsorship of the
Division of Industrial and Engineering Chemistry, Chicago, Illinois,
September 16 and 17, 1970

EDITED BY MILAN BIER
Veterans Administration Hospital and
The University of Arizona, Tucson, Arizona

ℚ PLENUM PRESS • NEW YORK–LONDON • 1971

Library of Congress Catalog Card Number 72-149647
ISBN-13: 978-1-4684-1913-9   e-ISBN-13: 978-1-4684-1911-5
DOI: 10.1007/978-1-4684-1911-5

© 1971 Plenum Press, New York
Softcover reprint of the hardcover 1st edition 1971

A Division of Plenum Publishing Corporation
227 West 17th Street, New York, New York 10011

United Kingdom edition published by Plenum Press, London
A Division of Plenum Publishing Company, Ltd.
Davis House (4th Floor), 8 Scrubs Lane, Harlesden, NW 10 6SE, England

PREFACE

The Symposium on Membrane Processes in Industry and Biomedicine has been held under the sponsorship of the Division of Industrial and Engineering Chemistry at the 160th National Meeting of the American Chemical Society, Chicago, Illinois, September 16 and 17, 1970. Its primary objective has been to spotlight some of the current directions of research in this rapidly growing field. There is at present considerable enthusiasm in membrane research, and the expectations are running high. This is partially due to the fact that basic concepts on which membrane processes are based are so deceptively simple. Moreover, all of us are living proofs of their potential efficiency. Our lungs and kidneys, skin and intestines are examples of membrane devices for gaseous and liquid separations, exchanges, and concentration. Even on a molecular level, life as we know is inconceivable without cell membranes and cell organs, such as mitochondria and chloroplasts, which appear to function as membrane regulated mini-factories for some of the most important and complex chemical syntheses in our bodies.

Membrane processes are not new. Experiments on gases were reported in 1831 by Mitchell. The origin of the whole field of colloid science is based on membrane studies, as in 1860 Graham discovered that some substances, for which he coined the name colloids, do not diffuse through parchment membranes, while most substances, termed crystalloids, readily do. Most enthusiastic worker in the field was probably the imposing Graff von Schwerin, who patented a whole series of processes in the 1910's and 20's, and established his Gesellschaft für Elektroosmose. No less ambitious were the aims of Pauli and Stamberger in the 1920's and 30's, who perfected electrodecantation to the point where Dunlop could operate huge membrane based latex creaming plants at some of their rubber plantations in Malaya.

The current renewed interest in membrane processes is largely due to advances in membrane synthesis, and reverse osmosis is the much heralded promised land. Its applications, however, are largely yet to be realized. The astonishing fact is that hemodialysis is probably the most successful single membrane process of today. Hemodialysis is, of course, the purification of blood through artificial kidneys. Over five thousand people in the world owe their lives to regular biweekly or even more frequent treatment with this membrane device. To anyone familiar with the complexity of functions handled by a normal human kidney, it is indeed amazing that such a simple membrane as cellophane or cuprophane could effectively substitute it. Not only by human value standards alone, but also by purely economic considerations the artificial kidney field is impressive as it has created a substantial new industry to serve the patients needs.

The papers are reproduced in this book in the same order as presented at the Symposium, which was divided into three sessions. The first session has dealt with artificial kidneys and other biomedically oriented devices. The second session was dedicated to reverse osmosis and heavy industry problems. The third was of interest to the biochemically oriented scientist, as it has dealt mainly with proteins and other biologicals.

The editor is much indebted to Mr. A. Fauver, Program Secretary, and Mr. A. R. Rescorla, Chairman of the Program Committee of the Division of Industrial and Engineering Chemistry, for making the Symposium possible. He also wishes to thank all the invited speakers for their cooperation, and particularly Drs. D. J. Lyman and A. S. Michaels, who have chaired two sessions. Finally, he acknowledges the editorial assistance of Mrs. C. Ragsdale.

Milan Bier
Symposium Organizer

LIST OF CONTRIBUTORS

G. L. Ball III, Monsanto Research Corporation, Station B, Dayton, Ohio

I. K. Bansal, The Institute of Paper Chemistry, Appleton, Wisconsin

G. L. Beemsterboer, Monsanto Research Corporation, Station B, Dayton, Ohio

M. Bier, Veterans Administration Hospital and University of Arizona, Tucson, Arizona

C. Calmon, Sybron Corporation, Cherry Hill, New Jersey

S. R. Caplan, Biophysical Laboratory, Harvard Medical School, Boston, Massachusetts

D. P. Carosella, Jr., Gulf General Atomic, San Diego, California

R. P. deFilippi, Abcor, Inc., Cambridge, Massachusetts

G. A. Dubey, The Institute of Paper Chemistry, Appleton, Wisconsin

H. Z. Friedlander, Union Carbide Research Institute, Tarrytown, New York

R. L. Goldsmith, Abcor, Inc., Cambridge, Massachusetts

A. H. Heit, Gamlen Chemical Company, San Francisco, California

S. Hossain, Abcor, Inc., Cambridge, Massachusetts

N. Lakshminarayanaiah, Department of Pharmacology, University of Pennsylvania School of Medicine, Philadelphia, Pennsylvania

F. B. Leitz, Ionics, Incorporated, Watertown, Massachusetts

N. N. Li, Esso Research and Engineering Company, Linden, New Jersey

L. M. Litz, Union Carbide Research Institute, Tarrytown, New York

H. K. Lonsdale, present address: Pharmetrics, Palo Alto, California

D. J. Lyman, Division of Materials Science and Engineering and Division of Artificial Organs, University of Utah, Salt Lake City, Utah

C. R. Lyons, Gulf General Atomic, San Diego, California

W. F. Mathewson, General Electric Company, Medical Development Operation, Schenectady, New York

W. A. McRae, Ionics, Incorporated, Watertown, Massachusetts

A. S. Michaels, present address: Pharmetrics, Palo Alto, California

L. Nelsen, Amicon Corporation, Lexington, Massachusetts

M. C. Porter, Amicon Corporation, Lexington, Massachusetts

R. L. Riley, Gulf General Atomic, San Diego, California
D. M. Ryon, General Electric Company, Medical Development Operation, Schenectady, New York
I. O. Salyer, Monsanto Research Corporation, Station B, Dayton, Ohio
F. A. Siddiqi, Department of Pharmacology, University of Pennsylvania School of Medicine, Philadelphia, Pennsylvania
R. S. Timmins, Abcor, Inc., Cambridge, Massachusetts
A. J. Wiley, The Institute of Paper Chemistry, Appleton, Wisconsin

CONTENTS

TRANSPORT PHENOMENA IN NATURAL AND SYNTHETIC MEMBRANES

S. Roy Caplan

Biophysical Laboratory, Harvard Medical School

Boston, Massachusetts

Although the differences between natural and synthetic membranes are as yet far more numerous than their similarities, the relationship of structure to function in certain instances is strikingly parallel. In principle there seems no reason why some of the specific functions of natural membranes should not eventually be emulated to great advantage on an engineering scale, and indeed there are encouraging indications that the initial steps have been taken in this direction. The most important general property of natural membranes is that they function as "active" elements (to use electrical network terminology). The transport of a given species is frequently <u>driven</u> by an input of metabolic energy, and in many cases the flow may be non-conservative—i.e., reaction and diffusion occur simultaneously within the membrane. In contrast, most synthetic membranes function as "passive" elements (this is true of the numerous types which have been used in purely physico-chemical studies, as well as those developed for specific purposes such as desalination, separations technology, or biomedical engineering). The permeability characteristics of these membranes are the only parameters of significance; no coupling to chemical reaction occurs and transport across them is always conservative. With one or two notable exceptions, moreover, they do not approach the extraordinary selectivity of natural membranes.

From several points of view the most exciting advance in the modelling of natural membranes has been the development of the synthetic lipid bilayer membrane (1). Lipid membranes are uni-

1

versally found in all living organisms. They are diverse in
character, and able to carry out (as well as compartmentalize)
fundamental activities essential for cellular life. The concept
of the lipid bilayer, or bimolecular leaflet, goes back to the work
of Gorter and Grendel on red blood cell lipids (2). They concluded
that the lipids of the red cell membrane were ordered in a double
layer, with the hydrocarbon tails facing inward toward each other
and the polar groups outward into the aqueous regions on either
side. Many studies have been carried out on synthetic membranes of
this kind over the last few years (3). A model which contains
lipid only oversimplifies the problem, however, since protein is
known to be just as important a structural component of biological
membranes as lipid. At present the most widely accepted scheme for
the molecular arrangement of lipids and proteins in the cell membrane is that proposed by Danielli and Davson (4), in which the
bimolecular lipid leaflet is covered on each face with a monolayer
of protein, the total thickness being about 80Å. These protein
layers, besides contributing to the stability of the membrane,
might be expected to confer on it specific transport properties.

Some indications of the extent to which the characteristics
of a lipid bilayer may be altered by the addition of certain molecules have already emerged from work with artificial bilayers.
Their electrical capacitances and water permeabilities are very
similar to the values found in cell membranes (5,6), but their
specific conductances and ionic permeabilities are less by a factor
of about a million (1,7). This difference vanishes when very low
concentrations of certain polypeptide or macrocyclic antibiotics,
such as gramicidin A or valinomycin, are present in the synthetic
bilayer (8,9). Furthermore, these substances confer a marked
selectivity on the membranes: for example valinomycin gives rise
to a potassium permeability nearly three hundred-fold greater than
that of sodium (8). Electrical excitability phenomena resembling
those found in nerve may also be exhibited under the influence of
certain complex proteinaceous mixtures (8,10).

These observations, together with many others, suggest that
the Danielli-Davson bimolecular leaflet structure may not be able
to account for the rich functional diversity of natural membranes.
For example, many enzymes are attached to membranes or subcellular
particles in vivo. These include the enzymes responsible for
active transport, the respiratory enzymes of mitochondria, and the
enzymes participating in photosynthesis and protein biosynthesis.
It is highly questionable whether such enzymes could function

solely as constituents of protein monolayers. Indeed it is now
well established that the cell membrane is a highly organized
dynamic structure in which many proteins and enzymes are located,
forming by their mutual interaction an essential part of the
regulatory machinery of the membrane (11).

MOSAIC MEMBRANES

Natural Mosaics

To many workers cellular membranes appear to be mosaics of
functional units formed by lipoprotein complexes. In this view the
proteins form the core of the complexes; phospholipids are attached
on the outside by van der Waals and coulombic forces and by hydrophobic bonds, while lipid layers may separate the complexes (12-
14). A similar view represents the membrane as a mosaic of globular lipid micelles in dynamic equilibrium with the bimolecular
leaflet structure: globular proteins with enzymatic properties
replace the lipid micelles at intervals in the plane of the lattice
(15,16). The membrane is penetrated by aqueous pores between the
micelles; these may be lined by the polar groups of the lipids or
by those of the protein molecules. The foregoing two views are
not necessarily mutually exclusive. In yet another view the lipid
micelles, extending between two protein layers, vary in shape from
cylindrical pillars (175-250Å in height) set rather far apart, to
closely abutting discs of bimolecular leaflet structure (17).
These two extremes represent "open" and "closed" configurations of
the membrane, and are believed to provide a basis, inter alia, for
active transport and impulse conduction (18).

The surface of a mosaic array of micelles corresponding to any
one or a combination of the above views is shown schematically in
Fig. 1. The array is not necessarily strictly hexagonal; individual micelles are in continuous although restricted random motion.
It has been suggested that those zones of the membrane having a
bimolecular leaflet structure may themselves be characterized by
mosaic patches of hydrophilic heads (of different phospholipids)
arranged so as to exhibit specificity toward a given protein (19).
Such a patch might "fit" a complementary grouping of charges on
the surface of the protein molecule.

Some of the most instructive examples of the functioning of

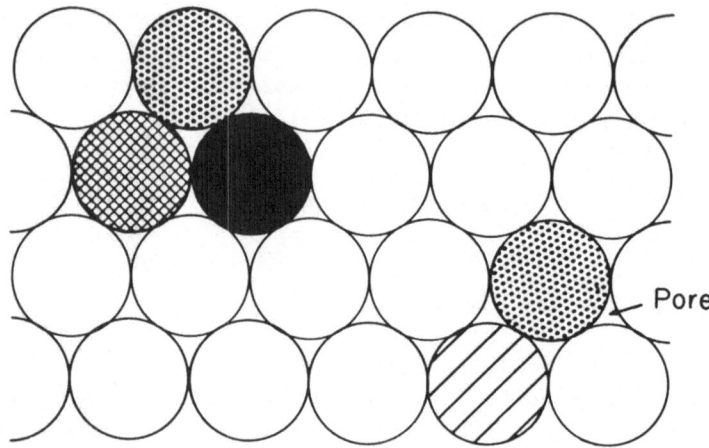

Fig. 1. Surface view of a micellar membrane: the mosaic array
represents lipid micelles interspersed with different globular
proteins or lipoprotein complexes. The micelles have over-all
diameters of about 50 Å, the aqueous pores effective radii of
about 4 Å. [After Lucy (16) and Kavanau (18).]

mosaic membranes are associated with the mitochondrion, the mem-
brane-enclosed organelle containing respiratory enzyme systems
found in the cytoplasm of aerobic cells. One such example relates
to NAD (nicotinamide adenine dinucleotide), the universal electron-
carrier in intracellular redox reactions. Cytoplasmic $NADH_2$ (the
reduced form of NAD) cannot be oxidized by intact mammalian mitochondria, since they are impermeable to it. However, a system for
the transfer of reducing power between cytoplasmic and mitochondrial
nicotinamide nucleotides has been described recently by Chappell
(20) and is depicted in Fig. 2. Four transport sites are involved,
two of which are subject to regulation. This system illustrates
the most typical property of mosaic membranes: they give rise to
circulatory flows. There are two circulatory pathways in Fig. 2—
the glutamate-aspartate circuit, and the malate-oxaloacetate-
oxoglutarate circuit. Circulation of this kind is intrinsic to the
rather special coupling phenomena characteristic of mosaics, a
striking illustration of which is Mitchell's chemiosmotic hypothesis

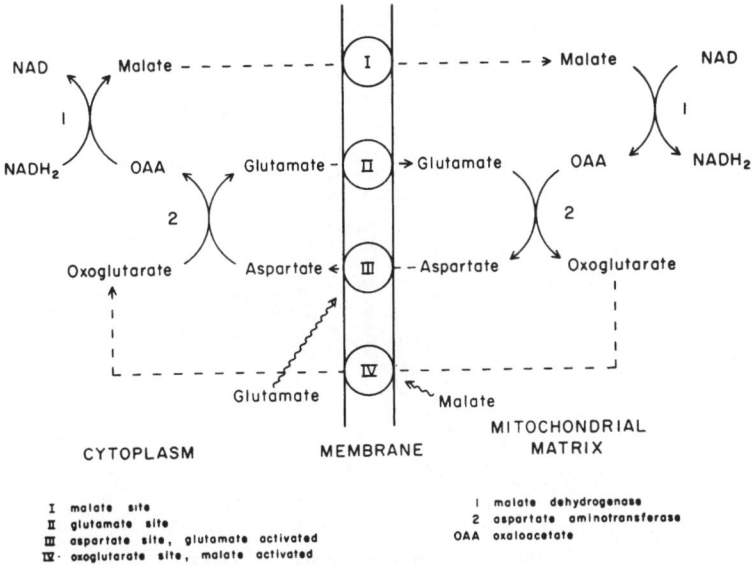

I   malate site
II   glutamate site
III aspartate site, glutamate activated
IV. oxoglutarate site, malate activated

1   malate dehydrogenase
2   aspartate aminotransferase
OAA oxaloacetate

Fig. 2. Transport system for the oxidation of cytoplasmic NADH₂ by mitochondria. [After Chappell (20).]

(21,22). This hypothesis for oxidative phosphorylation by mitochondria is shown diagramatically in Fig. 3 (23,24). Site I is equivalent to Mitchell's "proton-translocating oxido-reduction chain", site II to his "proton-translocating ATPase system". In this scheme electron transport through the respiratory chain drives the translocation of H^+, while the back flow of H^+ drives ATP synthesis. Oxidation and phosphorylation are coupled through the circulation of H^+, rather than through a possible high energy intermediate.

The analysis of transport phenomena in membranes built up of parallel and/or series arrays of elements is greatly facilitated by the application of nonequilibrium thermodynamics (35). Oxidative phosphorylation is a case in point: the analysis shows that the system has three degrees of freedom and hence can be represented by three "phenomenological" equations (24):

$$J_p = L_p A_p + L_{ph} \Delta\tilde{\mu}_h + L_{po} A_o,$$

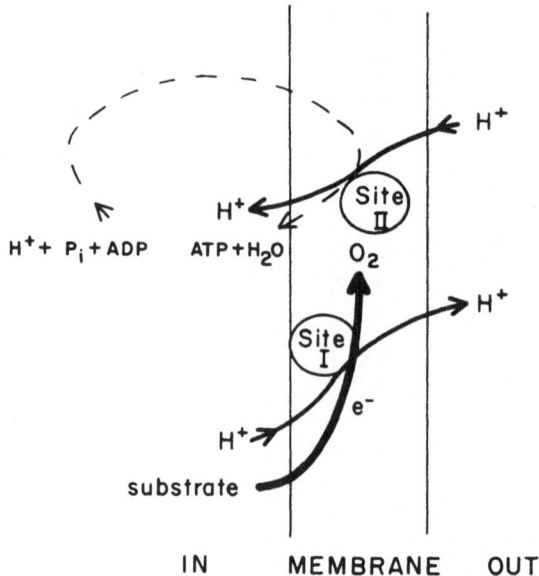

IN   MEMBRANE   OUT

Fig. 3. Oxidative phosphorylation: the chemiosmotic hypothesis. [After Mitchell (21,22) and Rottenberg et al. (23).]

$$J_h = L_{ph}A_p + L_h\Delta\tilde{\mu}_h + L_{oh}A_o,$$

$$J_o = L_{po}A_p + L_{oh}\Delta\tilde{\mu}_h + L_oA_o. \tag{1}$$

Here the J's are generalized flows, and the subscripts p, h, and o refer, respectively, to phosphorylation, net H^+ flow, and substrate oxidation. The A's are the affinities of the chemical reactions, ordinarily the negative free-energy changes per mole, measured in the extramitochondrial fluid. J_h is taken as the rate of decrease of H^+ content inside the mitochondrion, and the electrochemical potential difference $\Delta\tilde{\mu}_h$ is defined as $\tilde{\mu}_h^{in} - \tilde{\mu}_h^{ex}$. This general formalism should be applicable near equilibrium whether or not the chemiosmotic hypothesis is correct. For simplicity the usual assumptions of Onsager symmetry, i.e.

$$L_{ij} = L_{ji}, \tag{2}$$

and linearity have been made. According to the chemiosmotic

hypothesis, however,

$$L_{po} = 0, \tag{3}$$

i.e., if $\Delta\tilde{\mu}_h$ were maintained constant, phosphorylation and oxidation would be independent. In particular, if $\Delta\tilde{\mu}_h = 0$ uncoupling is complete, and $J_p = 0$ if $A_p = 0$, irrespective of J_o. This has recently been shown to be the case (38). If $\Delta\tilde{\mu}_h$ is not directly controlled, J_h becomes zero. Although neither remaining flow is now an explicit function of $\Delta\tilde{\mu}_h$, the existence of the circulatory H^+ transport mechanism influences the values of the phenomenological coefficients (39).

Impulse propagation in nerve may also be regarded as a mosaic phenomenon. In this case a transient mosaic structure which develops around the impulse permits local circulation of electric current. This occurs both in nonmyelinated and in myelinated nerve, as shown in Fig. 4. The resting potential of the axoplasm

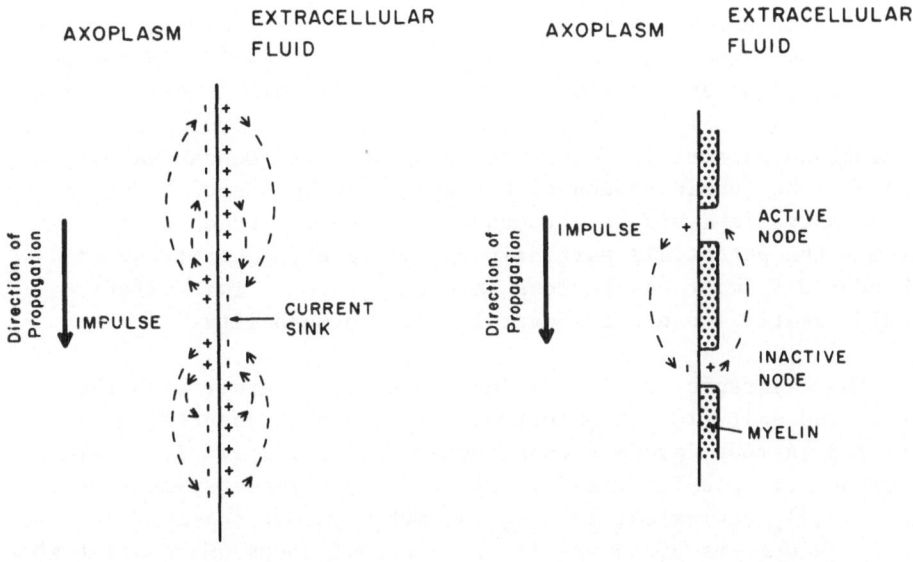

Fig. 4. Circulation of electric current during the propagation of a nerve impulse. [After Ganong (25).]

relative to the extracellular fluid, in the steady-state nerve
axon, is negative—due primarily to the unequal distribution of
ions maintained by a Na/K exchange pump, which utilizes metabolic
energy to expel sodium from the cell and accumulate potassium.
Since the resting axon membrane has a much higher permeability for
potassium than for sodium, the characteristic polarity is
established. During the passage of the impulse or action potential
the resting potential is first depolarized and then undergoes a
transient local reversal owing to a very large increase in sodium
permeability (which is potential dependent). Circulation of
current depolarizes the region or node ahead of the impulse causing
the sequence of events to be repeated (26). Thus the self-
propagating nature of the nerve impulse is intimately associated
with a circulatory phenomenon.

Synthetic Mosaics

    It was predicted by Sollner (27) that a mosaic membrane com-
posed of cation and anion selective elements (that is, a "charge-
mosaic") would exhibit unique transport properties arising from
electrical interaction between the unlike elements. He pointed
out that if such a membrane were placed between electrolyte
solutions of unequal concentration, circulating electric currents
should flow between the elements, as shown in Fig. 5. Homogeneous
ion exchange membranes are relatively impermeable to salts owing to
a more or less complete co-ion exclusion, and accordingly give rise
to membrane potentials which may approach ideal Donnan values.
However, the juxtaposition of homogeneous elements of opposite
fixed charge in a mosaic structure would be expected to short-
circuit the potentials partially or completely, permitting an
unhindered flow of counterions in each element. This effect
should greatly enhance the overall salt permeability.

    The existence of circulating currents, together with the
associated salt flux and electroosmotic water transport, was
verified in model systems constructed of two completely separate
membranes of opposite fixed charge (28-30). These systems were
functionally equivalent to mosaics, but were not characterized by
the large numbers of low-resistance current loops which arise when
many small elements are packed in a close array. The properties
of such arrays were recently described by Weinstein and Caplan
(31). They were prepared by embedding a single layer of
alternating cation and anion exchange beads in an impermeable

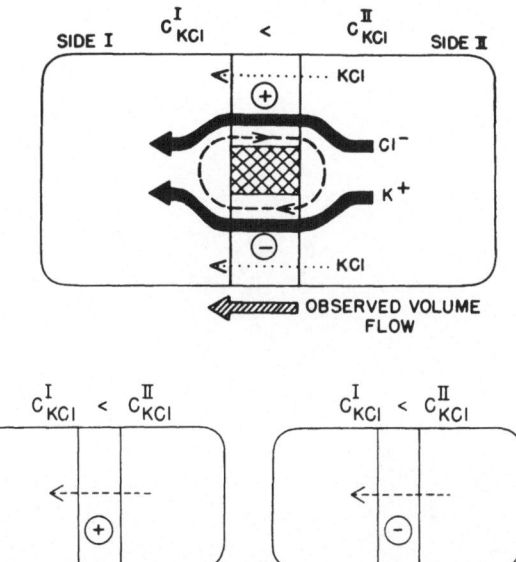

Fig. 5. Mechanism of permeability enhancement and negative osmosis in a charge-mosaic membrane. The upper diagram shows the circulating positive current brought about by juxtaposition of cation and anion selective elements (the fixed charges are indicated). The lower diagrams show the salt leakage in the elements when taken separately. [After Weinstein and Caplan (31).]

supporting matrix as shown in Fig. 6. Similar structures in which particles of anion exchange resin were embedded in a matrix with cation exchange properties have been described by Woermann (32).

Negative osmosis, in a system containing only one solute, may be defined as the occurrence of volume flow across a membrane from a solution of greater to one of lesser osmotic pressure, in the absence of hydrostatic pressure gradients and net flow of electric current across the membrane. With highly asymmetric electrolytes, such as K_2SO_4, $MgCl_2$, and strong acids, negative osmosis can take place across membranes having fixed charges of one sign only (33, 34). Although the nature of the mechanism is controversial, it is agreed that with such membranes negative osmosis occurs when the

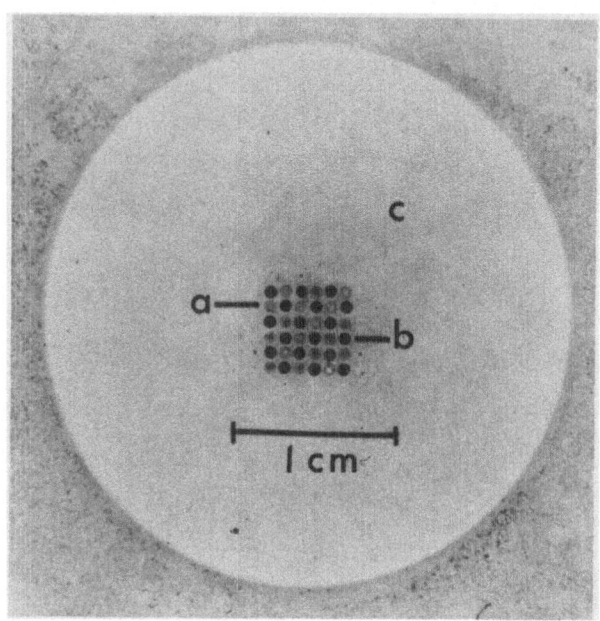

Fig. 6. A charge-mosaic membrane. The beads marked a and b are,
respectively, AG 1-X8 anion exchanger (exchange capacity dry
3.2 meq/g) and AG 50W-X8 cation exchanger (exchange capacity dry
5.1 meq/g) dyed with neutral red (both from Bio-Rad Laboratories,
Palo Alto, Calif.). The matrix material c is a flexible silicone
resin, Sylgard 184. [After Weinstein and Caplan (31).]

mobility of the co-ion is much greater than that of the counterion.
Prior to the work of Weinstein and Caplan (31) negative osmosis
had not been demonstrated in any type of membrane (homogeneous or
mosaic) with an electrolyte which is symmetrical in the sense that
the mobilities of both ions in aqueous solution are approximately
equal. However, the possibility that in charge-mosaic membranes
the electroosmotic transport of water due to circulating currents
might be large enough to overcome normal osmosis and produce
negative osmosis was anticipated by Sollner (27). This phenomenon
should not require an asymmetric electrolyte, as is also very
clearly demonstrated by the nonequilibrium thermodynamic calculations
of Kedem and Katchalsky (35).

In the formalism of nonequilibrium thermodynamics, the membrane parameter which determines the magnitude of the negative osmotic effect is the reflection coefficient σ (36,37), defined (for a system containing one solute) by

$$\sigma = \left(\frac{\Delta p}{\Delta \pi}\right)_{J_V=0, \ I=0} \tag{4}$$

where Δp and $\Delta \pi$ are the hydrostatic and osmotic pressure differences, respectively, J_V is the volume flow, and I is the net flow of electric current across the membrane. (Volume flow, J_V, and salt flow, J_S, are considered positive if they take place from side I to side II of the membrane, when $\Delta p = p^I - p^{II}$ and $\Delta \pi = \pi^I - \pi^{II}$.) For an ideally permselective membrane $\sigma=1$, whereas for a completely nonselective membrane $\sigma=0$. Negative osmosis occurs when $\sigma<0$. The reflection coefficient can also be expressed as (35)

$$\sigma = -\frac{1}{L_p}\left(\frac{J_V}{\Delta \pi}\right)_{\Delta p=0, \ I=0} \tag{5}$$

where L_p, the filtration coefficient, is defined by

$$L_p = \left(\frac{J_V}{\Delta p}\right)_{\Delta \pi=0, \ I=0} \ . \tag{6}$$

Since L_p is always positive, σ will be negative if J_V and $\Delta \pi$ have the same sign, that is, if volume flows toward the more dilute solution. The calculation of σ from the experimental results of Weinstein and Caplan is based on Eq. 5. The KCl permeability ω, defined by (35)

$$\omega = \left(\frac{J_s}{\Delta \pi}\right)_{J_V=0, \ I=0} \ , \tag{7}$$

was calculated from the measured values of J_s. As shown by the results in Table I the KCl permeabilities of the mosaics were 50 to 100 times those of the controls, indicating that salt "leakage" through the more or less permselective ion-exchange material accounted for only 1 to 2 percent of the over-all flux. The mosaics clearly exhibit pronounced negative osmosis.

It can be shown, on the basis of Onsager's reciprocal relations, that membranes capable of negative osmosis may in

TABLE I

EXPERIMENTALLY DETERMINED VALUES
OF σ AND ω

	$\omega \times 10^{15}$ (moles/dyne · sec)	σ
ANION EXCHANGE CONTROLS (3 expts.)	0.07 ± .02	+0.44 ± .02
CATION EXCHANGE CONTROLS (4 expts.)	0.11 ± .01	+0.89 ± .04
MOSAIC MEMBRANES (6 expts.)	5.1 ± 0.3	−1.10 ± .07

[After Weinstein and Caplan (31).]

principle be used for "piezodialysis," that is, extrusion through
the membrane of a salt solution which is more concentrated than
that on the high-pressure side. This process could possibly have
important applications in the field of desalination, but its
practicability has yet to be demonstrated.

SYNTHETIC ENZYMATIC MEMBRANES

A great deal of attention has recently been directed to the
function of enzymes when insolubilized by attachment to membranes.
Synthetic membranes containing bound enzymes have been shown to
yield important phenomena, relating inter alia to feed-back
regulation of chemical reaction, to facilitation of diffusion,
and to coupling of chemical reaction to flow. The first membrane
of this type to be described was prepared by absorbing papain in
a collodion membrane, and then cross-linking the papain in situ (40).

The pH-dependence of the activity of the enzyme membrane on a low
molecular-weight substrate, benzoylarginine ethyl ester, was found
to differ from that of crystalline papain; the activity was low in
the neutral pH range where the native enzyme has its optimum and
high at alkaline pH. This behavior was shown to be due to a
lowering of the local pH within the membrane as a result of the
release of acid by the enzymatic hydrolysis of the ester. The
data demonstrated that an enzyme embedded in a membrane can, by
its action, change its environment markedly and thereby its own
activity. More extensive studies on the papain-collodion membrane
showed that the effect could be obtained with other substrates,
and that the process of preparation resulted in a monomolecular
layer of papain within the pores of the collodion matrix (41).
Covalent binding of various enzymes to insoluble membrane matrices
has also been described by a number of workers (42,43).

The major effort in this area has focused on the kinetic
behavior of enzymatic membranes, a problem central to the understanding of the interaction between reaction and flow characteristic of these systems (44,45). It has been shown that if a substance undergoes simultaneous reaction and diffusion in a membrane
containing bound enzymes, the chemical production and mass transport can be uniquely resolved so as to satisfy a macroscopic
analogue of Curie's theorem, i.e. coupling between them is
impossible in a symmetrical membrane (46). Facilitated diffusion
may, however, be observed when a diffusing substance simultaneously
participates in a chemical reaction within a homogeneous membrane
(45).

If a chemical reaction is constrained to occur within an
asymmetric structure, on the other hand, coupling of the reaction
to the flow of one or more solutes, or to the flow of electric
current, becomes possible. Such systems can serve as models in
which transport is "driven" by chemical reaction. In this respect
the processes involved are analogous to active transport, though
the molecular mechanisms may be quite different from those in
nature. A simple arrangement of this kind has been studied: a
composite membrane consisting of two ion exchange membranes of
opposite fixed charge, separated by an intermediate layer of
solution containing papain (47). An uncharged amide substrate of
low molecular weight acted as "fuel" for the system, hydrolyzing
in the presence of papain to a salt. The composite membrane gave
rise to an electromotive force when clamped between two identical

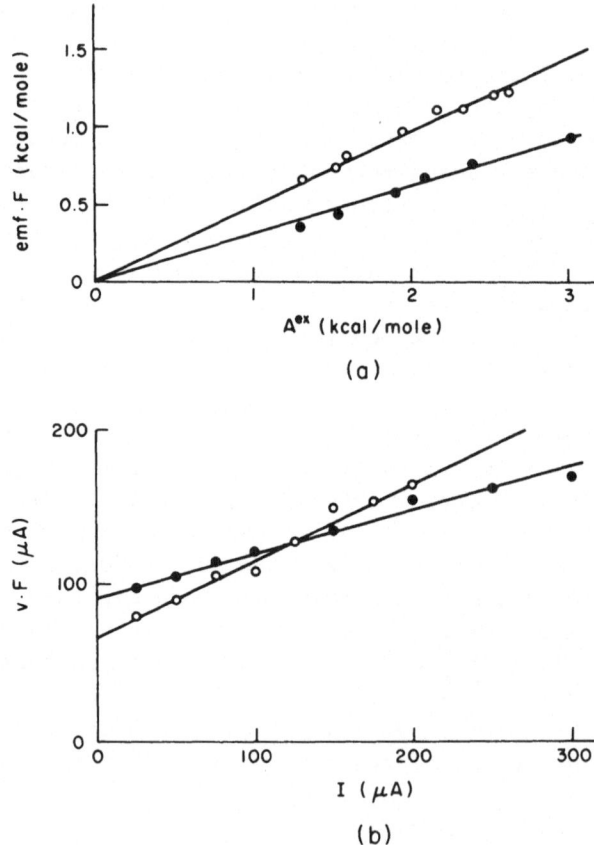

(a)

(b)

Fig. 7. Behavior of a model active transport system.
    (a) Steady-state electromotive force versus external reaction
affinity for two membrane systems. The experiments were performed
at zero current.
    (b) Steady-state reaction flow versus current for two mem-
brane systems. The experiments were performed at an affinity
A^{ex}=2.6 kcal/mole.
[After Blumenthal et al. (47).]

solutions in which the affinity of the reaction had been fixed. Onsager's reciprocity relation had not hitherto been tested in a case of coupling between chemical reaction and a vectorial flow (here electric current); its validity for this system was established over the experimental range of affinities (up to 3 kcal/ mole). The results of these model experiments have a strong bearing on the fundamental question as to whether biological transport processes can be characterized as <u>linear</u> functions, as can be seen by examining Fig. 7.

ACTIVE TRANSPORT

Active transport is perhaps one of the most characteristic functions of natural membranes, one of the most common species actively transported being the sodium ion. The literature of active transport is, however, fraught with confusion even as to its definition. A few authors have attempted to clarify matters by analyzing active transport from the standpoint of linear non-equilibrium thermodynamics—notably Kedem (48). In this framework the coupling between flow and reaction is defined unambiguously. Basing themselves on Kedem's analysis, Essig and Caplan treated the transport of a single cation driven by a single metabolic reaction (49). They showed that a composite system comprising a "pump", a series barrier, and a leak pathway may be described by linear equations, providing that each element shows linearity. A system of this kind is illustrated in Fig. 8. Two epithelial

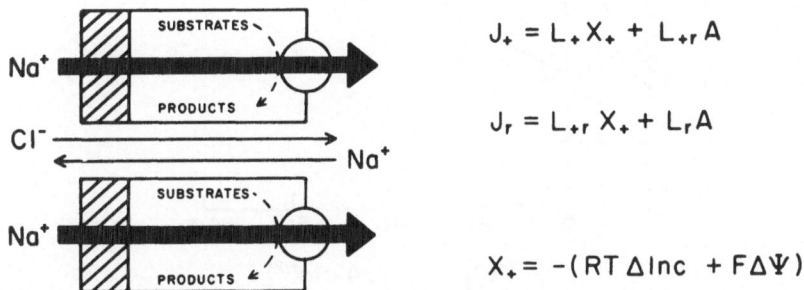

$$J_+ = L_+ X_+ + L_{+r} A$$

$$J_r = L_{+r} X_+ + L_r A$$

$$X_+ = -(RT \Delta \ln c + F \Delta \Psi)$$

Fig. 8. Thermodynamic description of an active transport system. J_+ and J_r represent the flow of sodium and the velocity of the coupled metabolic reaction, X_+ and A represent the electrochemical potential difference of sodium and the reaction affinity. [After Essig and Caplan (49).]

tissues which conform to this model in that they transport only
sodium actively are frog skin and toad bladder (50,51). For
simplicity chloride is considered to be representative of all
passively permeant anions. (The system is essentially a mosaic of
asymmetric cells and passive pores.) The affinity refers to some
region of the metabolic chain for which the pools of substrate and
product are large or maintained constant by the action of a
regulatory mechanism. The validity of this type of linear analysis
for epithelia may be judged from the recent work of Vieira et al.
on frog skin (52,53), as shown in Figs. 9 and 10.

CHEMODIALYSIS

    Biological processes have always played an important role
in the chemical process industries, and the handling of micro-

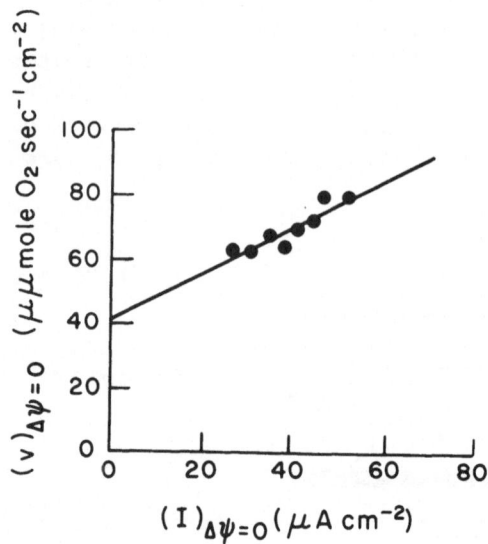

Fig. 9. The variation of short-circuit current and the associated
rate of oxygen consumption in a frog skin. [After Vieira et al.
(52).]

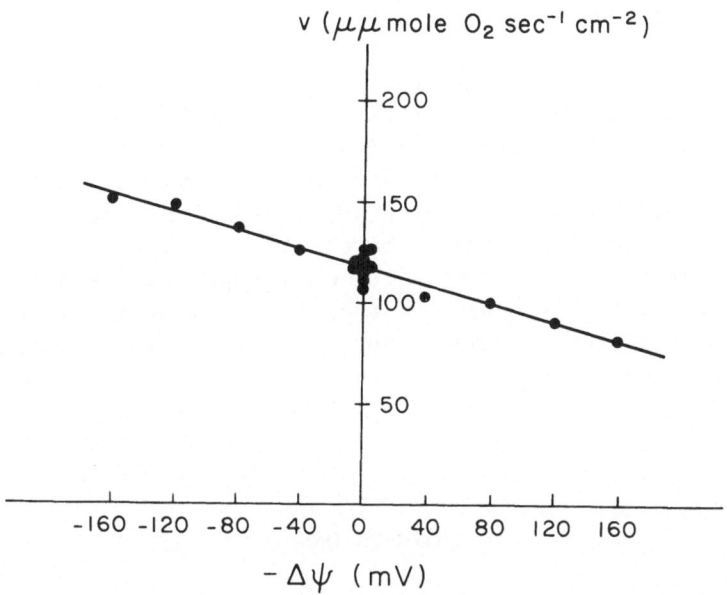

Fig. 10. Sodium transport in a frog skin: dependence of the rate
of oxygen consumption on the electrical potential difference $\Delta \psi$
in the presence of aldosterone ($\Delta \psi = \psi$ outside $^-\psi$ inside). [After
Vieira et al. (53).]

organisms and enzymes is an accepted part of chemical engineering.
Consequently the idea of utilizing a process akin to active transport for water desalination, using an organic "fuel" such as
crude sugar or alcohol, is hardly exceptional. Nevertheless no
serious exploration of this possibility, which may be termed
"chemodialysis," has yet been undertaken. The blame is obviously
to be placed on our meager understanding of the mechanism of
active transport. Although many examples of ion pumps in biology
have been studied in considerable detail, in no case has the mode
of action been elucidated. Plausible hypothetical models certainly exist, for instance the allosteric model of Jardetzky
(54) and the carrier protein model of Blumenthal and Kedem (55).
Since nature has in fact evolved a wide range of active transport
systems, there seems little doubt that synthetic membranes with
similar capabilities will ultimately be developed. Their structures,

however, are likely to require a high degree of orientation in the binding of the appropriate enzyme molecules to the appropriate polymer or lipid matrices.

A remarkable characteristic of biological ion pumps is their extreme selectivity, which in synthetic systems can only be matched by the antibiotic ionophores. In general such selectivity would not be required of a chemodialysis membrane, indeed it would prove disadvantageous. (One might conceive, however, of a membrane containing an array of pumps for different ions, some or all of which could be "switched on" by activation when required.) Since ion-pumps of low specificity would be entirely suitable for most desalination purposes, it may be possible eventually to synthesize chemodialysis membranes incorporating inorganic catalysts rather than enzymes, or organic "carriers" of relatively simple ion-exchange properties.

CONCLUSIONS

Natural membranes possess transport mechanisms of high specificity. These are frequently designed so that the transport is "active", i.e. coupled to metabolic chemical reactions and thus able to proceed against the electrochemical potential gradient of the transported species. An analogous process based on synthetic membranes and driven by a suitable fuel is conceivable; however, "chemodialysis" as a method for desalination or other separations may require the development of membranes having not only appropriate enzymatic activity but also relatively complex structures. The analysis of transport phenomena in membranes built up of parallel (mosaic) and/or series arrays of sub-units—some of which may incorporate enzymes—is greatly facilitated by the application of nonequilibrium thermodynamics. For example, it can be shown that the mosaic structures common to all natural membranes give rise to characteristic phenomena which may play an important role in such biological processes as oxidative phosphorylation; similar phenomena in synthetic mosaics are the basis of pressure dialysis and related techniques. At present the applicability of non-equilibrium thermodynamics is limited to the range of linearity between forces and flows. Nevertheless, in favorable circumstances this range can be remarkably wide in synthetic membranes, even when they are composed of arrays of sub-units: it can also be remarkably wide in natural membranes of great histological complexity.

ACKNOWLEDGMENTS

This work was supported by a grant from the Office of Saline Water, U.S. Dept. of the Interior (14-01-0001-2148), and by a U.S. Public Health Service Career Development Award (5 KO3 GM35292-04).

REFERENCES

1. Mueller, P., Rudin, D.O., Ti Tien, H., and Westcott, W.C. Nature 194:979 (1962).
2. Gorter, E., and Grendel, F.   J. Exp. Med.   41:439 (1925).
3. Tien, H.T., and Diana, A.L.   Chem. Phys. Lipids 2:55 (1968).
4. Danielli, J.F., and Davson, H.   J. Cell. Comp. Physiol. 5:495 (1935).
5. Hanai, T., Haydon, D.A., and Taylor, J.   Proc. Roy. Soc. A. 281:377 (1964).
6. Huang, C., and Thompson, T.E.   J. Mol. Biol. 15:539 (1966).
7. Hanai, T., Haydon, D.A., and Taylor, J.   J. Theoret. Biol. 9:433 (1965).
8. Mueller, P., and Rudin, D.O.   Biochem. Biophys. Res. Commun. 26:398 (1967).
9. Lev, A.A., and Buzhinsky, E.P.   Cytology 9:102 (1967).
10. Mueller, P., and Rudin, D.O.   Nature 213:603 (1967).
11. Nachmansohn, D.   Science 168:1059 (1970).
12. Benson, A.A.   J. Am. Oil Chem. Soc. 43:265 (1966).
13. Kennedy, E.P.   In The Neurosciences, G.C. Quarton, T. Melnechuk, and F.O. Schmitt, Eds.   Rockefeller University Press, New York, 1967, p. 271.
14. Rothfield, L. and Finkelstein, A.   Ann. Rev. Biochem.   37: 463 (1968).
15. Lucy, J.A.   J. Theoret. Biol. 7:360 (1964).
16. Lucy, J.A.   Brit. Med. Bull. 24:127 (1968).
17. Kavanau, J.L.   Nature 198:525 (1963).
18. Kavanau, J L.   Structure and Function in Biological Membranes, Vols. I and II.   Holden-Day, San Francisco, 1965.
19. Lehninger, A.L.   In Neurosciences Research Symposium Summaries, F.O. Schmitt and T. Melnechuk, Eds., Vol. I. M.I.T. Press, Cambridge, Mass., 1966, p. 294.
20. Chappell, J.B.   Brit. Med. Bull. 24:150 (1968).
21. Mitchell, P.   Nature 191:144 (1961).
22. Mitchell, P.   Biol. Rev. 41:445 (1966).

23. Rottenberg, H., Caplan, S.R., and Essig, A. Nature 216:610 (1967).

24. Caplan, S R., and Essig, A. Proc. Nat. Acad. Sci. 64:211 (1969).

25. Ganong, W.F. Review of Medical Physiology. Lange Medical Publications, Los Altos, 1963.

26. Hodgkin, A.L., and Huxley, A.F. Cold Spring Harbor Symp. Quant. Biol. 17:43 (1952).

27. Sollner, K. Biochem. Z. 244:370 (1932).

28. Neihof, R., and Sollner, K. J. Phys. Colloid Chem. 54:157 (1950).

29. Neihof, R., and Sollner, K. J. Gen. Physiol. 38:613 (1955).

30. Carr, C., and Sollner, K. Biophys. J. 4:189 (1964).

31. Weinstein, J.N., and Caplan, S.R. Science 161:70 (1968).

32. Woermann, D. Ber. Bunsengesell. Physik. Chem. 71:87 (1967).

33. Grim, E., and Sollner, K. J. Gen. Physiol. 40:887 (1957).

34. Grim, E., and Sollner, K. J. Gen. Physiol. 44:381 (1960).

35. Kedem, O., and Katchalsky, A. Trans. Faraday Soc. 59:1918, 1931, 1941 (1963).

36. Staverman, A.J. Rec. Trav. Chim. 70:344 (1951).

37. Staverman, A.J. Trans. Faraday Soc. 48:176 (1952).

38. Rottenberg, H. European J. Biochem., in press.

39. Katchalsky, A., and Spangler, R. Quart. Rev. Biophys. 1:127 (1968).

40. Goldman, R., Silman, H.I., Caplan, S.R., Kedem, O., and Katchalski, E. Science 150:758 (1965).

41. Goldman, R., Kedem, O., Silman, H.I., Caplan, S.R., and Katchalski, E. Biochemistry 7:486 (1968).

42. Mitz, M.A. In Proceedings of the Conference on Natural and Synthetic Membranes, Boston, 1967, C. Saravis et al., Eds. U.S.D.H.E.W., Bethesda, p. 208.

43. Selegny, E., Avrameas, S., Broun, G., and Thomas, D. C. R. Acad. Sci. Paris 266c:1431 (1968).

44. Goldman, R., Kedem, O., and Katchalski, E. Biochemistry 7:4518 (1968).

45. DeSimone, J.A., Owen, A., and Caplan, S.R. Biophysical Soc. Abstracts 10:71a (1970).

46. DeSimone, J.A. and Caplan, S.R. In Mass Transfer in Biological Systems, A.L. Shrier and T.G. Kaufmann, Eds. Chem. Eng. Progr. Symp. Ser. 66:43 (1970).

47. Blumenthal, R., Caplan, S.R., and Kedem, O. Biophysical J. 7:735 (1967).

48. Kedem, O. In Membrane Transport and Metabolism, A. Kleinzeller and A. Kotyk, Eds. Academic Press, New York, 1961, p. 87.

49. Essig, A., and Caplan, S.R. Biophys. J. 8:1434 (1968).

50. Ussing, H.H., and Zerahn, K. Acta Physiol. Scand. 23:110 (1951).

51. Leaf, A., Anderson, J., and Page, L.B. J. Gen. Physiol. 41:657 (1958).

52. Vieira, F.L., Caplan, S.R., and Essig, A. In preparation.

53. Vieira, F.L., Caplan, S.R., and Essig, A. In preparation.

54. Jardetzky, O. Nature 211:969 (1966).

55. Blumenthal, R., and Kedem, O. Biophys. J. 9:432 (1969).

PARTITION MEMBRANES FOR AQUEOUS DIALYSIS

Donald J. Lyman

Division of Materials Science & Engineering and Division

of Artificial Organs, University of Utah, Salt Lake City

Utah   84112

One of the purposes of this symposium is to focus on the
state-of-the-art of a problem area - hopefully, to stimulate new
ideas and approaches. Of interest in this section is the area of
membranes for artificial kidney devices.

During this last decade, we have seen an almost spectacular
development in the use of an artificial organ to maintain life,
from about six chronic kidney failure patients in 1960 to over
three-thousand in the United States alone in 1970[*]. It would seem
that little else is needed except to supply more devices and training so that all that need treatment would have it available. However, much more is needed. If we are to develop the full potential
of these artificial kidney devices for all types of acute kidney
failure treatment, including regional perfusion, chronic kidney
failure, and research, we need new membrane materials. To put this
need for new biomaterials in proper perspective, it is of interest
to briefly review the development of the artificial kidney.

In 1913, Abel, Rowntree and Turner[1] built a device designed to
dialyze blood of experimental animals under sterile conditions.
While these hemodialysis experiments were essentially successful,
the lack of good and reproducible membranes and anticoagulants
kept this from becoming anything more than a laboratory experiment.
It was not until thirty years later, when the availability of both

---

*On the basis of current medical selection criteria, it is estimated
that there are about 7,500 suitable candidates/year for such treatment.

the anticoagulant heparin and a relatively inexpensive reproducible
cellulose membrane material (i.e., Cellophane), did we see the
hemodialyzer move from the laboratory to the clinic.   In 1943,
Dr. Willem Kolff developed a rotating drum type of "artificial kidney," and used it successfully in a patient with acute kidney
failure[2].

During the next ten to fifteen years, there were many engineering and medical refinements of these devices by Alwall, Skeggs
and Leonards, MacNeill, Kiil, and others[3-7]. This work led, in
the late 1950's, to two major types of devices--the coil artificial
kidney and the flat sheet artificial kidney.

During all of these years, the artificial kidney was used
primarily for patients with acute kidney failure. Since restora-
tion of renal function was anticipated, usually only one to two
dialyses were needed. Extensive repetitive dialysis, such as would
be required by a chronic kidney failure patient, was not considered
possible. This restriction was essentially based on access to the
blood system. The shortage of suitable radial arteries and veins,
and their subsequent loss of accessibility after the surgical procedure needed for dialysis, meant that only a limited number of
dialyses could be performed. The longest patient survival using
such intermittant dialysis was 181 days[8].

In the late 1940's, Alwall experimented with an indwelling vas-
cular cannula by-pass system as a technique to provide easy access
to the patient's blood system[9]. However, only glass and rubber
tubings were available for his by-pass device, but these clotted
severely when the patient was continuously heparinized. It was
not until 1960, with the commercial availability of polytetrafluoro-
ethylene (Teflon) and polydimethyl siloxane (Silastic rubber), that
Scribner and Quinton were able to develop a practical indwelling
cannulae system for prolonged hemodialysis[10]. With this development began the era of prolonged chronic dialysis. From the original few patients in 1960 (three are still alive), we have seen
the procedure expand until today it is an accepted clinical treatment keeping approximately five thousand patients alive in the world.

Improved engineering during this decade has greatly reduced
the cost of artificial kidney equipment, has improved the controls
and monitoring of the devices so that dialysis can be done by the
patient in his home, and in general has made this clinical technique available to a large number of people. New technology, such
as hollow fibers, has been utilized in the development of novel
kidney designs. One of these will be discussed in the following
paper. However, even with these tremendous accomplishments during
this last decade, again we are on a plateau. New materials are
needed if we are to advance the potential of dialysis.

The current cellulosic membrane materials are essentially the
same as that used by Kolff in the 1940's. While we have been able
to save many lives, recent data indicate not only a disturbing
trend in less full rehabilitation of the surviving patients, but
an increase in patient complications, especially in the second and
third year of treatment[11]. For example, full rehabilitation has
decreased from 56% in 1965 to 43% in 1969, and those not rehabili-
tated have increased from 26% to 34%. We have also seen increases
in nephrogenic neuropathy, pseudo-gout, metastatic calcification,
secondary hyperparathyroidism with bone disease, psychosis, etc.
Many nephrologists believe that insufficient removal of medium and
large size molecules, such as uric acid, and possibly materials up
to several thousand molecular weight, may be responsible for these
developments of secondary complications[12]. Several groups, including our own[13,14], have explored techniques to improve the dialysis
of larger molecules. One approach is to increase the pores of
cellulose membranes by a variety of techniques, such as biaxial
stretching and chemical etching of the membrane, modified regenera-
tion processes, etc. While membranes having increased porosity
can be made, the procedures often lack good reproducibility. In
addition, several problems, such as a general weakening of the membrane, a leaking of protein, and too great an ultrafiltration of
water, can be serious limitations to this approach.

A second approach would be to create new types of dialysis
membranes which separate molecules primarily on the basis of chemical structure rather than physical size. Such membranes might
act, for example, in an aqueous system like Silastic rubber membranes do in a gaseous system. i.e., involve a solubility factor as
well as a diffusion factor in their transport mechanism.

In designing the molecular structure of polymers that might
be suitable for such permeable selective membranes, one is drawn
to consider the block copolymer systems. In these structures, a
block, or macrosegment, would be of such hydrophilicity that it
would form a water soluble area. The other areas would be hydrophobic. The former would contribute water swelling and possible
selective partitioning of the solute into the membrane; the latter
would contribute to the mechanical strength of the water swollen
membrane. The morphological organization of the three dimensional
mosaic structure formed by these block copolymers would also contribute to the overall properties. Work on these "partition"
membranes was pioneered in our laboratory[15-17].

Initial studies were concerned with copolyether-ester structure
based on polyoxyethylene glycol and polyethylene terephthalate.

$$\left\{ \overset{O}{\overset{\|}{C}} \langle \bigcirc \rangle \overset{O}{\overset{\|}{C}} O(CH_2CH_2O)_n \overset{O}{\overset{\|}{C}} \langle \bigcirc \rangle \overset{O}{\overset{\|}{C}} OCH_2CH_2O \right\}_n$$

These were chosen because of their relative ease of synthesis. Of
particular interest was the copolymer containing 30 mole percent
of polyoxyethylene glycol, molecular weight 1500. This polymer
was cast into clear, tough films which retained their strength when
swollen in water. Determination of the half-time rates of transfer of various solutes through this membrane showed no clear-cut
relationship between molecular weight and diffusion time (see
Table I and Figure I). This is in contrast to Cuprophane membranes
in which diffusion is through discrete pores and does show a rather
linear relationship between molecular weight and diffusion time.

TABLE I. Half-time transfer rate ratios of Cuprophane/Copolyester
          membrane.

Less than 0.91	Approx. 1.0 (0.94 - 1.07)	Greater than 1.10
Glycine (0.82)	Glutamine (1.0)	Urea (1.73)
L-Alanine (0.85)	D,L-Phenylala- nine (0.96)	Creatinine (1.48)
β-Alanine (0.54)	Glucose (1.07)	D-Threonine (1.25)
Sarcosine (0.72)	Sucrose (0.94)	L-Leucine (1.32)
D,L-Serine (0.91)		Histidine (1.32)
L-Proline (0.87)		Uric Acid (1.96)
L-Valine (0.89)		Arginine (1.29)
Hydroxyl-L-Pro- line (0.76)		Ascorbic Acid (1.79)
Raffinose . $5H_2O$ (0.62)		Citric Acid (1.14)
		Glucuronic Acid (1.41)
		Glucosamine · HCl (1.34)
		Thiamine Chloride (1.36)
		Bacitracine (1.81)

(left margin, vertical: Increasing Mol. Wt.)

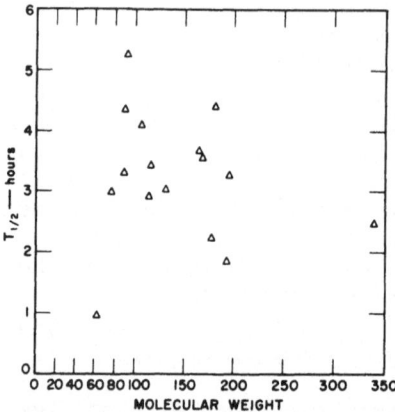

FIGURE 1. The effect of molecular weight on the half-time rates of
transfer (T-1/2) of various compounds through a copolyether-ester
membrane. Reprinted from Trans. Amer. Soc. Artif. Int. Organs (ref.
15).

Determination of distribution coefficients for various compounds between water and copolyether-ester membrane indicates that, in many cases, the unique separation possible with these membranes can be explained on the basis of partitioning. For example, L-alanine and β-alanine, which dialyze through Cuprophane in an identical manner can be separated using the copolyether-ester membrane. L-alanine (T-1/2 and 201 minutes) has a distribution coefficient of 0.52, while β-alanine (T-1/2 of 317 minutes) has a distribution coefficient of 0.25. This appears to be the first example of membranes for aqueous dialysis in which solute-membrane interactions are sufficient to influence transport rates.

A variety of block copolymer systems were then explored to determine the effect of variations in the structure of the hydrophilic and hydrophobic segments. (For example, hydrophilic segments such as polyethers, polypeptides, polyvinyl alcolhol, polyacrylic acid, etc., hydrophobic segments such as urethanes, esters, amides, hydrocarbons, etc.) Of particular interest in our current work are the copolyether-urethanes, since within one generic family a variety of membranes with unusual properties can be prepared. These membranes also have greater hydrolytic stability than the polyester type.

The copolyurethanes are best prepared by a two-step process using solution polymerization techniques[18,19].

$$NO(CH_2CH_2O)_xH \;+\; 2\; OCN\text{—}\bigcirc\text{—}CH_2\text{—}\bigcirc\text{—}NCO \;\rightarrow$$

$$OCN\text{—}\bigcirc\text{—}CH_2\text{—}\bigcirc\text{—}HN\overset{O}{\overset{\|}{C}}O(CH_2CH_2O)_x\overset{O}{\overset{\|}{C}}NH\text{—}\bigcirc\text{—}CH_2\text{—}\bigcirc\text{—}NCO$$

I

$$0.3\; I \;+\; 0.7\; HO(CH_2)_5OH \;+\; 0.4\; OCN\text{—}\bigcirc\text{—}CH_2\text{—}\bigcirc\text{—}NCO \;\rightarrow$$

$$O\overset{O}{\overset{\|}{C}}N\text{—}\bigcirc\text{—}CH_2\text{—}\bigcirc\text{—}NH\overset{O}{\overset{\|}{C}}O(CH_2CH_2O)_x\overset{O}{\overset{\|}{C}}NH\text{—}\bigcirc\text{—}CH_2\text{—}\bigcirc\text{—}NH\overset{O}{\overset{\|}{C}}\text{-}O(CH_2)_5O\text{-}$$

$$\left[\overset{O}{\overset{\|}{C}}NH\text{—}\bigcirc\text{—}CH_2\text{—}\bigcirc\text{—}NH\overset{O}{\overset{\|}{C}}O(CH_2)_5O\right]_n$$

II

   In this series the molar ratio and molecular weight of the
polyoxyethylene glycol were kept at constant; the type of diol
used to form the hydrophobic segment was varied. Surprisingly, it
was found that relatively minor changes in the hydrophobic segment,
resulting from changing the diol coupler, led to a great variance
in membrane diffusion properties. For example, as shown in Figure
2, changing the diol from cis/trans 1,4-cyclohexane-diol (I) to
1,10 decanediol (V), one can increase urea transport, while greatly
reducing the transport of the other solutes. By changing to 1,5-
pentanediol (III), a separation between α- and β-alanine occurs.
Use of other diols also gives variation in solute transport through
these copolyether-urethane membranes. These structural changes
mainly effect the configuration of the hydrophobic segment (and
as a result the overall morphology in the phase domains). Therefore it would appear that we have a greater latitude in designing
these partitioning membrane, since we can vary both the hydrophilic
and hydrophobic segments. As a result, it should be possible to
tailor a membrane for any end-use application.

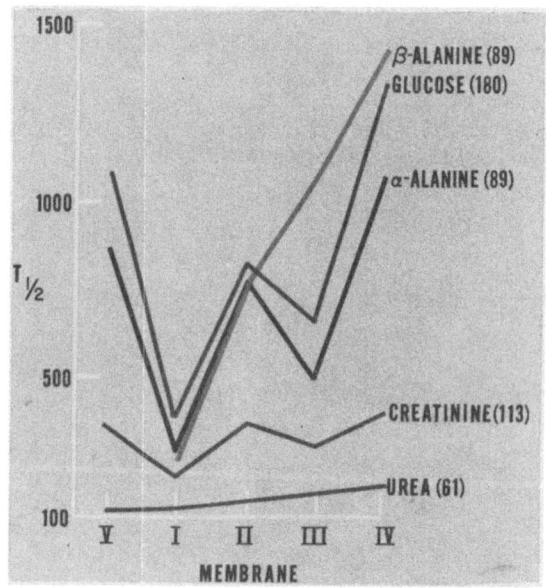

FIGURE 2. The relative half-time rates of transport (Minutes) of
various solutes through copolyether-urethane membranes (based on
polyoxyethylene glycol) in which the structure of the hydrophobic
urethane segment is varied. The diol couplers are: membrane I,
cis/trans-1.4-cyclohexanediol; membrane II, trans-1,4-cyclohexanediol;
membrane III, 1,5-pentanediol; membrane IV, 1,10-decanediol; membrane
V, α-α'-dihydroxy-p-xylene. Reprinted from ref. 20.

This brings us to several additional problems faced by the membrane researcher. First, what solutes do we want the membrane to effectively remove from the blood? In acute poisoning situations this is easy to determine. However, in chronic kidney failure we do not know what must be removed from these patients (nor what should be retained). Second, how do we determine the clinical potential of a new membrane? We can determine the removal rates of urea, creatinine, uric acid, water, etc., but what might happen to the patient after several years of dialysis? We have seen the onset of "secondary diseases" from improper dialysis by cellophane membranes.

Research is progressing in both of these areas and hopefully, in the not too distant future, we will have the answers. In the meantime, we are studying the diffusion of selected solutes through these new partition membranes to determine the mechanism of transport and to relate this mechanism to the molecular structure of both solute and membrane polymers.

In these studies[21] the true resistance for mass transfer of a membrane is determined and is related to the interactions between solute and membrane, and between solute and water. By using a porous membrane (such as a Millipore filter) as a control, it is possible to relate these factors as follows:

$$\frac{\overline{D}_2}{\overline{D}_1} = \frac{\delta_2 \Phi \tau}{K_D \delta_1} \left(1 \pm \frac{U_{sm}}{U_{sw}}\right)$$

where $\overline{D}_1$, δ_1, Φ, τ are the membrane diffusion coefficient, wet thickness, porosity, and tortuosity factors for the Millipore filter; $\overline{D}_2$, δ_2, K_D, are the membrane diffusion coefficient, wet thickness, and distribution coefficient for the homogeneous polymer membrane; U_{sm} and U_{sw} are the solute-membrane and solute-water resistances. The relationship of the $\overline{D}_2/\overline{D}_1$ ratio to the $\delta_2 \Phi \tau / K_D \delta_1$ value gives an indication of the degree of solute-membrane interaction. Experimental results do indicate that solutes pass through Cuprophane membranes in a manner similar to that shown by the Millipore filter. On the other hand, the copolyether-urethane membranes show a high degree of selectivity.

Detailed studies on the nature of the distribution coefficient are in progress. These studies, coupled with the diffusion studies on a series of copolyether-urethanes (in which the molecular weight, molar concentration, and segmental arrangement of the polyoxyethylene glycol, and the structure of the hydrophobic segment are varied), should lead to a general understanding of these new partition membranes.

References

1. J. J. Abel, L. G. Rowntree, and B. B. Turner, J. Pharmacol. and Exper. Therap., 5, 275 (1914).

2. W. J. Kolff and H. T. J. Berk, Geneesk Gids., 5, 21 (1943).

3. N. Alwall, Acta. Med. Scand., 128, 317 (1943).

4. L. T. Skeggs and J. R. Leonards, Science, 108, 212 (1948).

5. A. E. MacNeill, J. E. Doyle, R. Anthons, S. Anthone, New York State J. Med., 59, 4137 (1959).

6. F. Kiil, Acta. Churg. Scand. (Suppl.), 253, 142 (1960).

7. W. J. Kolff and J. Van Noordwijk, De Kunst. Natige Nier., Kampden, Holland, J. H. Kok Publisher, 1946.

8. J. E. Doyle, Extracorporeal Hemodialysis Therapy in Blood Chemistry Disorders, C. C. Thomas, Springfield, Ill., 1962.

9. N. Alwall, L. Norvitt, and A. M. Steins, Acta. Med. Scand., 132, 587 (1949).

10. W. Quinton, D. Dillard, and B. H. Scribner, Trans. Amer. Soc. Artif. Int. Organs., 6, 104 (1960).

11. H. Klinkmann to D. J. Lyman (1970) private communication.

12. H. Klinkmann and W. J. Kolff, Present Concepts Inter. Med., 3, 437 (1970).

13. D. J. Lyman, Ann. N. Y. Acad. Sci., 146, 113 (1968).

14. H. Klinkmann, D. J. Lyman, R. L. Kirkham, V. K. Kulshrestha, and W. J. Kolff, Trans. Amer. Soc. Artif. Int. Organs., 16 121 (1970).

15. D. J. Lyman, B. H. Loo, and R. W. Crawford, Biochem., 3, 985 (1964).

16. D. J. Lyman, Trans. Amer. Soc. Artif. Int. Organs, 10, 17 (1964).

17. D. J. Lyman, B. H. Loo, and W. M. Muir, Trans. Amer. Soc. Artif. Int. Organs, 11, 91 (1965).

18. D. J. Lyman and B. H. Loo, J. Biomed. Mater. Res., 1, 17 (1967).

19. D. J. Lyman, J. Polymer Sci., 45, 49 (1960).

20. D. J. Lyman, Excerpta Medica Intern. Cong. Series, 155, 98
     (1967). (4th Conference E.D.T.A., Paris, June 1967).

21. D. J. Lyman, S. I. Yum, R. Haske.., and M. Yoon, manuscript
     in preparation.

THE MONSANTO POLYACRYLONITRILE HOLLOW FIBER ARTIFICIAL KIDNEY

I. O. Salyer, G. L. Ball III, G. L. Beemsterboer

Monsanto Research Corporation

SUMMARY

Advantages of a hollow-fiber hemodialyzer have been well described.[1-3] The goal of this program was to provide a working model of a noncellulosic hollow fiber artificial kidney with non-thrombogenic blood-contacting surfaces, low blood-priming volume and low pressure drop, a satisfactory rate of urea transport, and high rate of ultrafiltration of water. Also, the reliability and utility of these devices was to be demonstrated through clinical evaluation.

To define materials for hollow fibers, preliminary studies were conducted with various membrane compositions including polyvinyl alcohol, ethylene vinyl alcohol copolymers, ethylene/acrylate co-polymers, nylon, and polyacrylonitrile (PAN). Membranes of modified PAN gave urea resistance values, *in vitro*, of 15 min/cm. *In vivo* testing, by medical collaborators at Peter Bent Brigham Hospital, Boston, Mass.,gave resistance values of 17 min/cm, which compared to a value of 19 min/cm for Bemberg Cuprophane. Ultrafiltration rates could be regulated up to 300 ml/min/m^2 using a transmembrane pressure differential of only 25 mm Hg.

Using the results of our membrane research and our background knowledge of the spinnability of various polymeric materials, a modified PAN was selected as the prime candidate for development as a dialyzing hollow fiber. This material could readily be rendered nonthrombogenic by quaternizing groups in the polymer followed by ionic coupling with heparin.

Three multiple-bundle, parallel-flow dialyzing subscale artificial kidneys incorporating the modified PAN hollow fibers were prepared and tested *in vitro* and *in vivo*. Urea transport of the fiber was significantly lower than that observed in the membrane. However, the high rate of water ultrafiltration and the nonthrombogenic properties were confirmed.

To obtain greater simplicity and lower cost 1/6-scale <u>single bundle dialyzers</u> were then designed, fabricated, and tested. The units consisted of approximately 2000 modified PAN fibers having inside diameters of ∿300 microns, and 15-cm dialyzing length, which were encased in a disposable cartridge with nonthrombogenic blood-contacting surfaces. This sterilizable, disposable cartridge design is low in cost and lends itself to the ease of handling and operation necessary for patients on home dialysis.

Two of these single-bundle, hollow fiber prototype artificial kidney units were tested *in vivo* in patients suffering from chronic renal failure by Drs. Merrill, Hampers and Lowrie at Peter Bent Brigham Hospital. These units were used independent of any other dialyzers, and demonstrated that the modified PAN hollow fiber kidneys

were truly nonthrombogenic,

provided a permeability of 0.01 to 0.015 cm/min (urea clearance of 34 ml/min),

exhibited very high ultrafiltration (which was adjustable over any desired clinical range), and

were of clinical utility.

Scale-up of the current design indicates that a full-size kidney of present fiber composition would require ∿12,000 fibers. With foreseeable improvements in the fiber dialysis characteristics, as indicated from membrane data, a practical full-scale hemodialyzer could evolve from the prototype 2000-fiber unit.

EXPERIMENTAL RATIONALE

Selection and Evaluation of Candidate Polymeric Membranes

The principal polymer systems examined in the film form as dialysis membranes were amine-modified polyacrylonitrile (PAN), polyurethanes, hydrolyzed ethylene/vinyl acetate copolymers, hydrolyzed ethylene/vinyl pyridine terpolymer, nylon 66, nylon 4, polyvinyl alcohol, and hydrolyzed ethylene/alkyl-acrylate copolymers. These membranes were formed using various casting techniques, rates of film formation, film thicknesses, and drying rates.

The membranes were examined using standard dialysis static cells and dynamic type Babb and Grimsrud cells[4],[5]. Standard burst and tear tests (ASTM D-1004-66) were also run.[6]

The amine-modified PAN film provided the best combination of dialysis and strength properties. The amine could be quaternized and ionically coupled with heparin to provide thromboresistance.

Urea resistance values of 15 min/cm were obtained *in vitro*. Urea resistance values obtained *in vivo* were as low as 17 min/cm compared to 19 min/cm for Bemberg Cuprophane. The burst strengths of the modified PAN membranes ranged from 3 to 4 pounds per mil thickness. Very high ultrafiltration rates up to 300 ml water/min/ m^2 were also observed.

This membrane material was selected as the material for use in hollow fiber kidneys because it combined both ease of spinnability from a solvent system and ease of being rendered nonthrombogenic through heparinization, with very good permeability and ultrafiltration rates.

The other polymer systems were found to be lacking in terms of spinnability, availability of nonthrombogenic character, mechanical strength, or poor permeability.[7]

Development and Evaluation of Amine-Modified Polyacrylonitrile Hollow Fibers

Based on the findings from the membrane analysis and our background knowledge of the spinnability of various materials, hollow fibers, spun from amine-modified PAN, were examined for dialysis, ultrafiltration, and nonthrombogenicity. These hollow fibers were spun under various conditions to determine the effect of spinning parameters on these characteristics.

The modified PAN fibers were spun with inside diameters of 200 to 300 microns and wall thicknesses of 30 to 100 microns.

As a means for determining the dialysis and ultrafiltration characteristics of the hollow fibers, a technique was developed for combining about four hundred fibers into bundles approximately 65 millimeters long. The technique involved potting the fiber ends and combining them into a cylindrical shape with shrink-down tubing and plastisol.

In vitro characterization of these bundles of PAN hollow fibers showed that urea transport depended very strongly on the spinning conditions and especially on the "structure" formed during coagulation.

It was also shown that the effect of this "structure" was so pronounced that the thick-walled fibers actually had a higher rate of permeability than comparable thin-walled ones.

The best fiber systems showed blood urea permeabilities between 0.008 and 0.015 cm/min and ultrafiltration (water) of about 11 ml/min/m^2 at 45 mm Hg transmembrane pressure. Although these permeability rates were promising, the values were only approximately 20% of those that had been obtained in an optimum film membrane of the same composition. Thus by varying the coagulation and spinning conditions, fibers having transport rates comparable to the best PAN membranes could probably be obtained.

Hollow Fiber *In Vivo* Characterization Dialysis Devices

Design and Assembly. Multiple-module dialysis units were designed and assembled. The devices consisted of seventeen to twenty modules in which each module contained approximately four hundred modified PAN fibers. These devices provided dialyzing units for *in vitro* and *in vivo* characterization of experimental fibers.

The prime advantages of this approach were:

> The ability to test the efficacy of the individual modules of experimental fibers *in vitro* and thus to then evaluate only qualified fibers *in vivo*.

> The ease of fabrication of a small number of fibers at one time.

> The ease of fabrication of short fiber lengths, which reduced blood flow resistances.

> The adaptability of this approach to mixed experimental fibers to provide specific dialysis and ultrafiltration conditions.

The disadvantages of the multiple-module unit were the numerous joint areas that had to be sealed under nonthrombogenic conditions and the less-than-good potential packing efficiency.

The enclosure cases used for the dialysis units were constructed of polymethylmethacrylate. One of the units was coated with a heparinized epoxy in all areas that would contact blood. The multiple module characterization dialysis device therefore was completely nonthrombogenic.

The case designed for the multiple module unit is illustrated
in Figure 1 and in Figure 2. The case is transparent and consists
of two primary sections: (a) the dialyzate section and (b) the
blood manifold. The dialyzate section is the portion of the case
in which the outer surface of the hollow fibers are exposed and
through which dialyzate is pumped. The blood section comprises
the blood manifolds which feed the bores of the hollow fibers through
which blood is pumped.

Characterization. The physical parameters and performance of
these dialyzing multiple-module characterization units are shown in
Table 1. Unit 3 represented the state-of-the-art of the experimental
fibers prior to the design and assembly of a dialyzing artificial
kidney.

The efficacy of a multiple module dialysis device as an *in vivo*
characterization tool was also determined. Unit 3 was prepared from
the best modified PAN fibers using the best fabrication techniques.
Problems encountered on previous models were eliminated in the production of Unit 3.

The ultrafiltration of the amine-modified PAN fibers was measured *in vitro* on quaternized but unheparinized single 400 channel
modules. The ultrafiltration rate for each of these bundles was 1
ml/min at a transmembrane pressure differential of 280 mm Hg. Therefore, the total ultrafiltration rate for Unit 3 was 20 ml/min or
about 1 liter per hour.

Unit 3 was evaluated *in vivo*. The *in vivo* evaluation was conducted by placing the unit in parallel with the Kolff twin-coil
kidney during dialysis of a patient. At various times during the
dialysis, blood samples were removed from the arterial and venous
sides of the unit.

The results obtained in the *in vivo* dialysis are shown in Table
2. They reflect attempts to manipulate the outflow pressure to increase the ultrafiltration rate. The ultrafiltration rate itself,
however, was not measured.

The last three columns of Table 2 show the membrane permeability,
resistance, and urea dialysis as calculated from the *in vivo* results.
These data, coupled with the ultrafiltration conducted *in vitro*,
indicated a relatively high rate of urea transport which nearly
approximated that of cellophane.

Following the dialysis characterization, Unit 3 was examined
for the presence of clots. Only a few fibers clotted; this was
attributed to flattening of those fibers. Some of the fibers ruptured, but were sealed at once by strings of clots. The rupture of
the fibers, however, was due to inadvertent over-pressuring of the

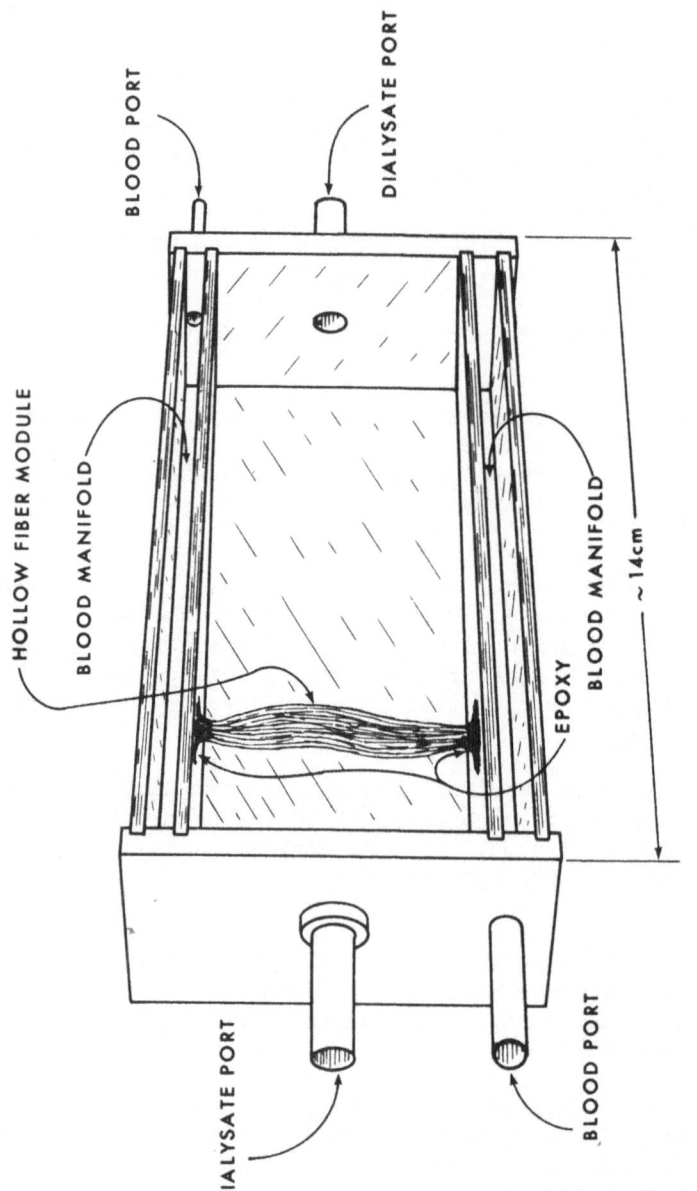

Figure 1.   Schematic Drawing of Prototype, Hollow Fiber, *In Vivo* Characterization, Dialysis Device.

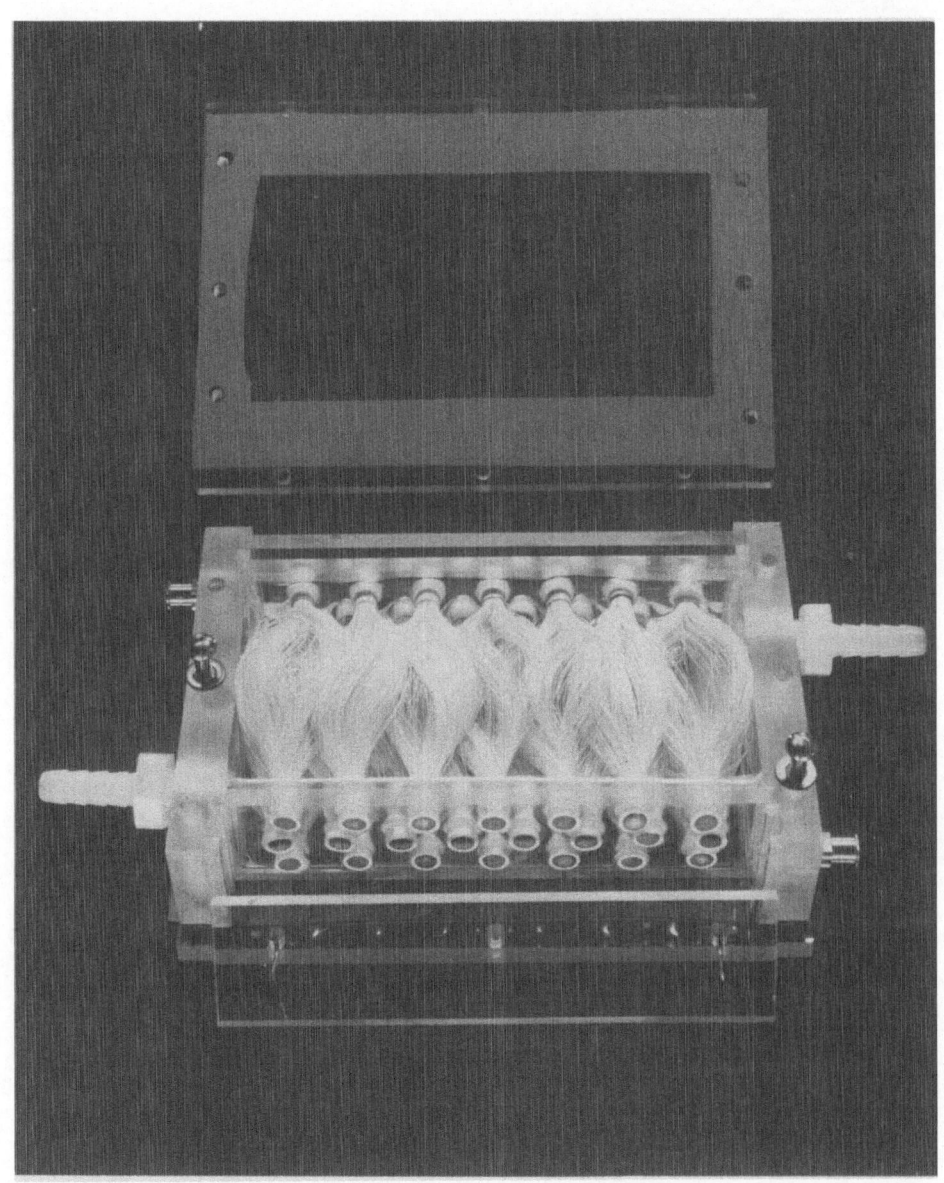

Figure 2.   Hollow Fiber, Multiple-Module, *In Vivo*-Characterization,
Dialysis Device.

Table 1

IN VITRO CHARACTERISTICS OF THE MODIFIED
POLYACRYLONITRILE MULTIPLE-MODULE DIALYSIS UNITS

	Multiple Bundle Dialysis Units		
	#1	#2	#3
Fiber Bore, microns	200	200	230-250
Fiber OD, microns	370	400	350-370
Total Fiber Channels	5,100	8,000	8,000
Effective Channel Length[1], cm	5	5	5
Total Internal Dialyzing Area[2], square meter	0.16	0.25	0.30
Water Flow @ 50 mm Hg ml/min	170	800	800
Total Blood Flow (Calculated) @ 50 mm Hg, ml/min	34	160	160
Fibers Quaternized and Heparinized	No	Yes[3]	Yes
Case Heparinized	No	No	Yes
Submitted Hospital Evaluation, Date	No	No[4]	5/13/69

---

[1]Actual fiber length 7 cm
[2]Based on inside diameter of fibers
[3]Quaternized only
[4]Broken and leaky fibers

Table 2

IN VIVO DIALYSIS PERFORMANCE OF THE MULTIPLE-MODULE,
MODIFIED PAN HOLLOW-FIBER CHARACTERIZATION UNIT 3[1]

Blood Sample	Pressure[2] Inlet (mm Hg)	Outlet (mm Hg)	B.T. Blood (sec)	Urea Conc. Arterio (mg, %)	Venous (mg, %)	Creatinine Conc. Arterio (mg, %)	Venous (mg, %)	Membrane Permeability (cm/min x 10^{-4})	Membrane Resistance (min/cm)	Urea Dialysance (ml/min)
1	30	105	3.1	48	41	8.45	7.7	101	99	29
2	10	109	4.0	47	37	8.35	7.3	119	84	31
3	0	115	3.9	43	36	8.15	7.0	91	110	25
4	10	174	4.0	39	32	7.40	6.55	100	112	26
Average								100	100	28

[1] Conducted 16 June 1968 by C. L. Hampers, M.D.,
Peter Bent Brigham Hospital, Boston
#3 unit contains heparinized modified PAN fibers.
Dialyzing area = 0.3 square meters

[2] Blood flow rate = ∿150 ml/min
Dialysate flow rate = ∿1200 ml/min

system to about 400 mm Hg pressure. There was some minor fibrin collection around the header ports, which although definitely present did not pose any significant problem.

Prototype, Single-Bundle, Hollow-fiber Artificial Kidney

Requirements and Design. Various designs for a total dialysis artificial kidney device were examined. It was concluded that one of the most desirable approaches was to prepare a single bundle of a large number of hollow fibers and to incorporate them into a case.

The advantages of a single-bundle cartridge artificial-kidney (SBC) dialysis-unit were:

A desirable packing efficiency and thus greater surface area-to-volume ratio for the system.

The few seals and seal areas that must be nonthrombogenic.

The ease of imparting nonthrombogenicity to the blood-contacting portions of the case.

The continuity and homogenity of the single potting required.

The ready availability of a throw-away dialysis cartridge design.

The potentially economic cylindrical case arrangement.

The disadvantages of the single bundle artificial kidney dialysis unit were:

The increased difficulty of initially combining and holding the numerous fibers.

The initial problem of potting and achieving a totally leak-free potted seal with numerous fibers.

The total acceptability or rejectibility of a single unit due to the failure of even just one fiber (which probably could be overcome by improved fibers, inspection, and repair techniques).

A single-bundle artificial kidney prepared from fibers of a specific bore diameter and dialysis characteristic can be designed in a number of configurations. To provide for a practical consideration of the design of our hollow fiber artificial kidney, it was therefore necessary that a number of boundary conditions be defined.

These consisted of the following:

A dialysance of 125 ml/min.

An arterio-venous pressure of 50 mm Hg.

Blood flow of 200 ml/min.

A minimum blood priming volume.

Blood viscosity of 0.0347 poise in tubing and 0.025 poise in capillaries.

Nonthrombogenicity of all blood-contacting surfaces.

A dialyzate flow of at least 500 ml/min.

Light weight (portable and storable).

Low cost per dialysis (considered disposable)

Sterilizibility of blood-contacting areas.

Operability with minimum training of personnel.

Compatibility with the state-of-the-art dialyzate and cannula systems.

Based on the above boundary conditions and available fibers, the number and length of fibers necessary to produce a full-scale dialysis unit was calculated. The fibers available from our development work had a permeability of 0.010 to 0.015, inside diameters of 0.025 cm, and wall thicknesses of 0.005 cm (average).

The requirements for the design of the prototype single-bundle artificial kidney involved combining about 2000 fibers of 15 cm dialyzing length into a bundle through which blood could be passed and around which dialyzate would flow.

The amine-modified PAN fibers with inside diameters of 250 to 280 microns and wall thicknesses of about 50 microns were used. This wall thickness was selected as a compromise between dialysis performance and burst strength, but with the existing fibers represented a 1/6-scale unit.

We therefore considered a completely disposable artificial kidney. It became apparent very soon, however, that a case for these fibers could not readily be prepared that would be disposable on an economic basis. Accordingly, we designed a case that would be a permanent item and included in it a cartridge of hollow fibers which itself was disposable. This arrangement may be likened to the automobile oil filter cartridge.

The design that evolved is illustrated in Figures 3 and 4. The disposable cartridge, shown in the center of Figure 4, consists of a polymethylmethacrylate cylinder into which the 2000 fibers were potted. This cartridge had holes drilled into its side at the ends of the unpotted region to allow for passage of the dialyzate around the outside of the fibers.

The case for the cartridge, shown in the top of the photograph, was designed so that the cartridge fit snugly with only one O-ring seal to provide separation of the dialyzate inlet and outlet. The dialyzate is fed into the cartridge through the ports on the side of the case, which connects to an annular channel around the inside circumference of the case. The cartridge is the same length as the case.

To provide for blood entry to the inside of the fibers all that is required is clamping of the nonthrombogenic headers and seals to both ends of the case by means of stainless steel end fixtures.

The assembled hollow fiber artificial kidney is shown in Figure 5, and a view demonstrating insertion of the disposable cartridge is shown in Figure 6.

A solution for one of the most important performance require- ments of this artificial kidney was provided through new and advanced materials. This requirement was that all blood-contacting surfaces be nonthrombogenic. Nonthrombogenicity was provided by fabricating all the blood-contacting surfaces, including the blood headers, seals, and fiber potting materials from a nonthrombogenic epoxy. The tech- niques for preparing this thromboresistant epoxy were developed for the National Institutes of Health, the National Heart and Lung Institute.[8,9,10] The nonthrombogenic character of a modified poly- acrylonitrile has already been discussed.

In Vitro Evaluation. Single-bundle units were built and tested for dialysis and ultrafiltration *in vitro* with the apparatus shown in Figure 7. These units contained the best fibers at the time and because of their dialysis characteristics were found to be about 1/16 the actual size necessary for a total dialysis unit.

A variety of single-bundle units were made. The characteristics of these various units are listed in Table 3. The most important characteristics of the modified PAN fibers were the demonstrated high ultrafiltration rates and acceptable dialysis available. The ultra- filtration and dialysis characteristics of these fibers are compared to other systems in Table 4. The results on the SBC prototypes were

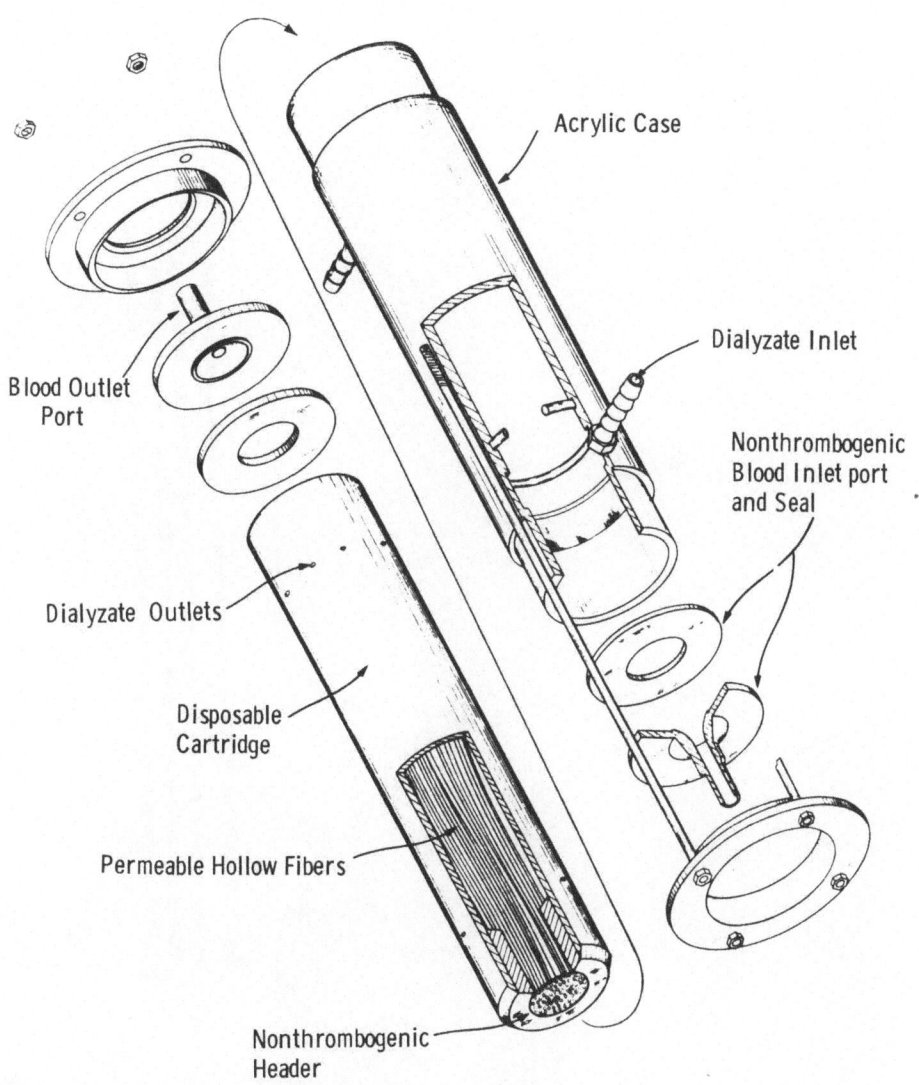

Figure 3.   Exploded View of Prototype, Hollow-Fiber, Artificial-
Kidney with Disposable Dialyzing Cartridge.

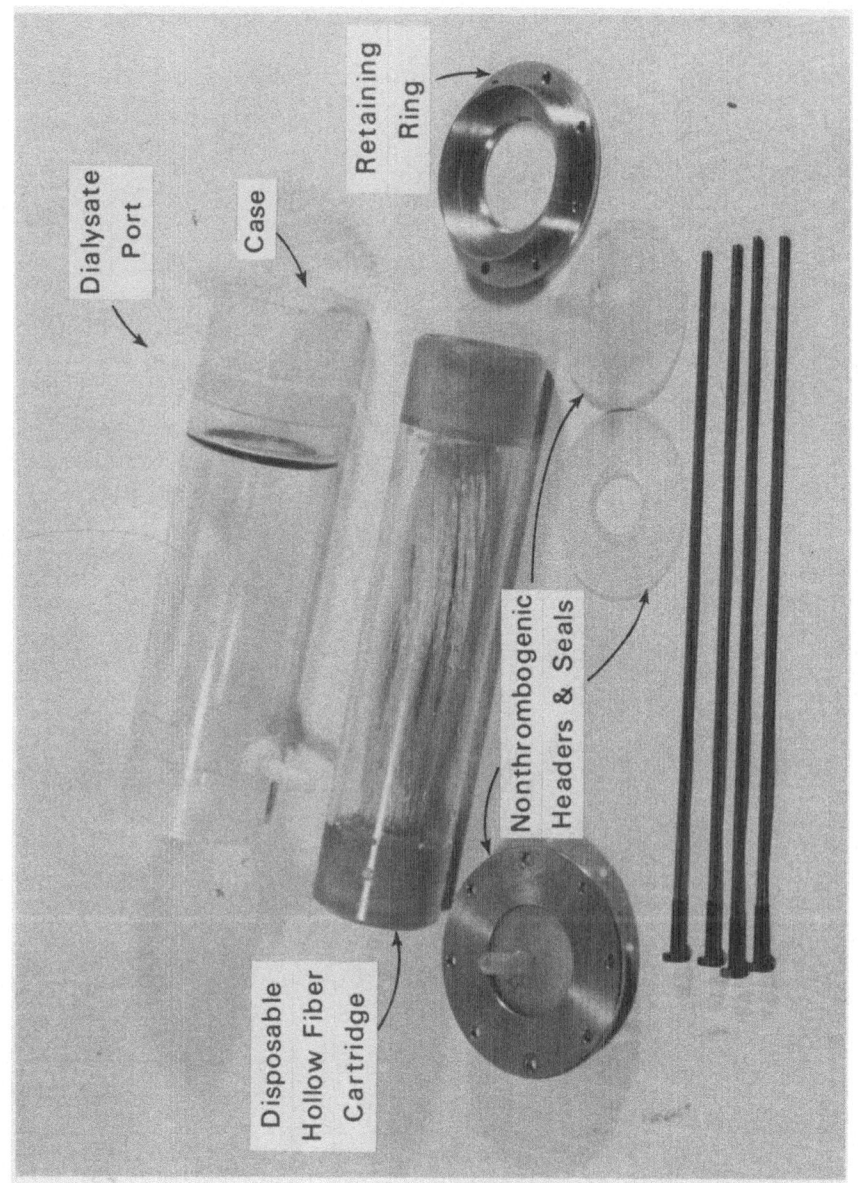

Figure 4.   Dissassembled PAN, Hollow Fiber, Prototype Kidney.

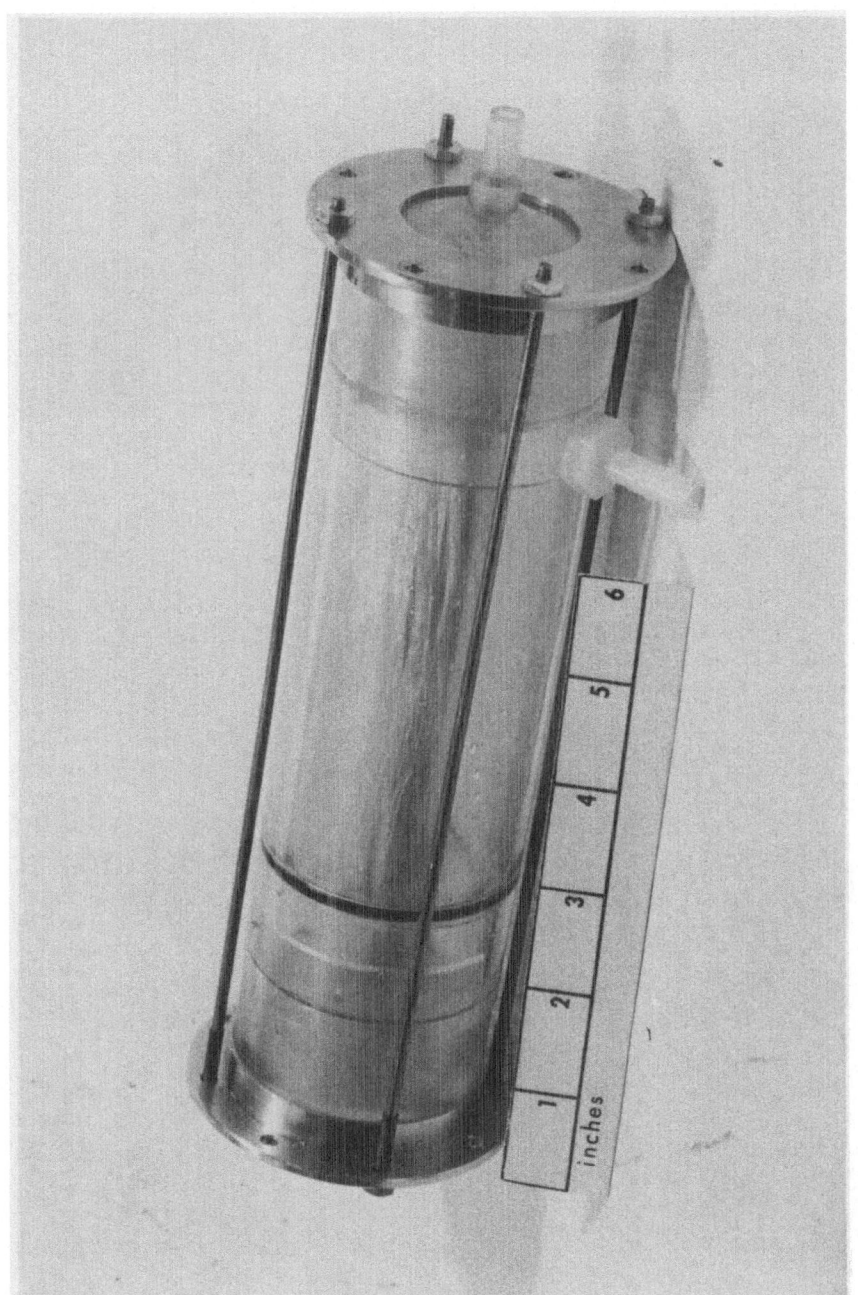

Figure 5. Physical-Prototype, Artificial Kidney Containing Modified Polyacrylonitrile Hollow Fibers.

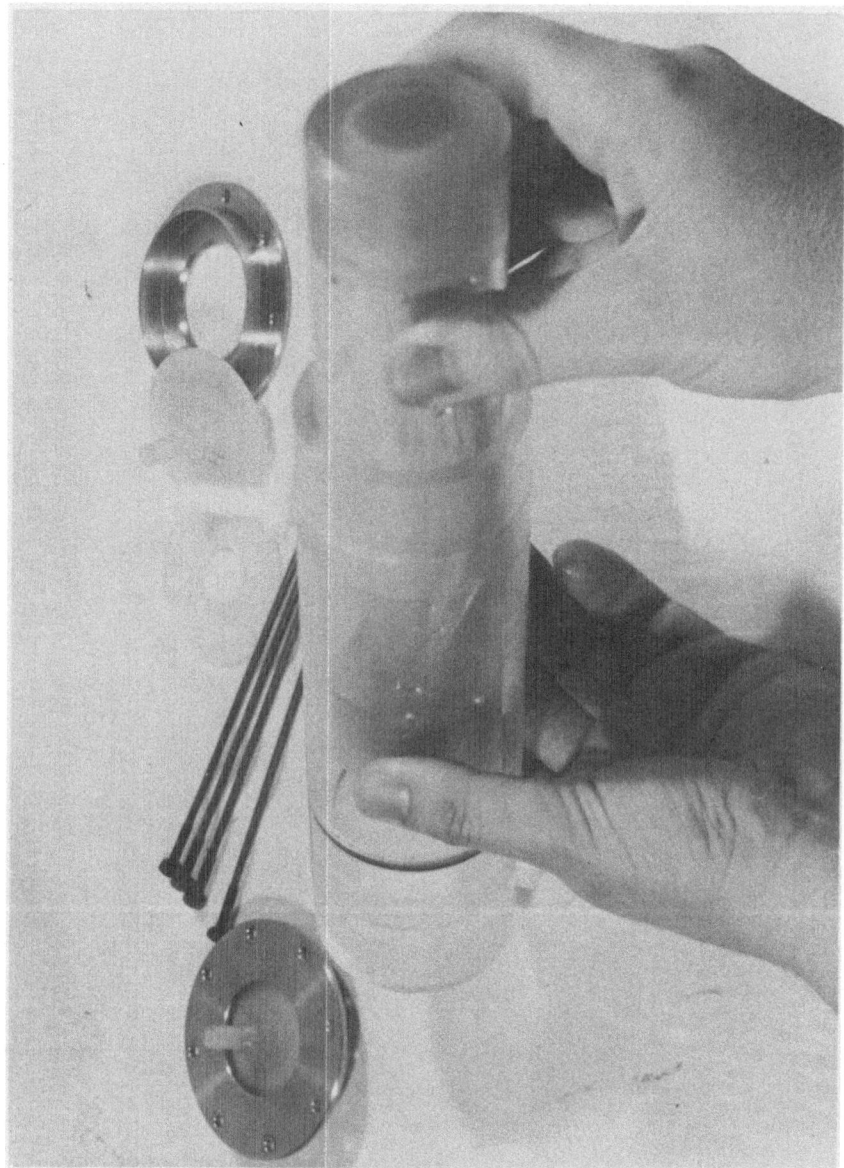

Figure 6.   View of PAN Hollow Fiber, Prototype Kidney Showing Insertion of the Disposable Cartridge.

Figure 7. *In Vitro* Dialysis and Ultrafiltration Test Arrangement for the Hollow Fiber Prototype Artificial Kidney.

Table 3

IN VITRO CHARACTERISTICS OF THE PROTOTYPE
SINGLE-BUNDLE, HOLLOW-FIBER DIALYSIS CARTRIDGES

	Single Bundle Cartridge			
	SBC-1	SBC-2	SBC-3	SBC-4
Fiber[1] Bore, Microns	250-280	250-280	250-280	250-280
Fiber O.D., Microns	370-400	370-400	350-370	350-370
Total Channels	2000	1850	1850	2000
Effective Unit Channel Length, cm	15	15	15	15
Total Internal Dialyzing Area[2], sq m	0.24	0.22	0.22	0.24
Total Blood Flow[3], @ 50 mm Hg, ml/min	160	150	165	160
Fibers Heparinized	No	Yes	Yes	Yes
Case Heparinized	No	Yes	No	Yes
Clinical Evaluation	-	7/29/69	-	8/8/69

[1]Quaternized modified polyacrylonitrile
[2]Based on fiber bore diameter
[3]Calculated from module scale tests on H_2O flow

Table 4

DIALYSIS AND ULTRAFILTRATION CHARACTERISTICS OF THE MODIFIED PAN FIBERS[1] AND THE PROTOTYPE SINGLE-BUNDLE CARTRIDGES

Identification	Fiber Inside Diameter (microns)	Fiber Wall Thickness (microns)	Area (cm²)	Membrane Resistance (min/cm)	Permeability Coefficient (cm/min, x 10⁻⁴)	Ultrafiltration Rate (ml/min/m²)	Transmembrane[2] Pressure (mm Hg)
Fibers[3]	250[4]	~50	185	165 150 300	60 67 33	12 35 100	45 99 280
Fibers	250	~60	185	100	100	50	280
Multiple Module (Unit 3)	250	~60	~3000	100[5]	100	-	300-400
SBC-2 Prototype[6]	250	~60	~2200	100	100	50	280
SBC-3 Prototype[7]	250	~60	~2200	135 67	126 152	0 23	54 100
SBC-4 Prototype[8]	250	~50	~2350	67	152	23	100
Membrane[9]	-	100	90	10-205	500-1000	300	25
Cellophane Membrane[10]	-	100	90	33	300	5	25

1 Quaternized and heparinized
2 Independent Variable
3 Quaternized only
4 Diameter Range 250-280
5 Measured *in vivo* at Peter Bent Brigham Hospital
6 Data extrapolated from quaternized and heparinized fibers *in vitro*
7 Data determined empirically *in vitro*
8 Data extrapolated from SBC-3
9 Data calculated from Babb-Grimsrud results
10 Data entered for base line comparison

derived using the data from SBC-3. The fibers and materials were basically the same and this derivation resulted from the use of *in vitro* data on smaller bundles of fibers.

Very high dialysis and ultrafiltration rates were potentially available in this system (based on membranes) but because of the fiber characteristics, the membrane performance could not be fully translated to the fibers.

In Vivo Evaluation. Two of our complete nonthrombogenic single-bundle prototypes, prepared using modified PAN fibers, were evaluated *in vivo*. This evaluation was conducted in the dialysis center of the Peter Brigham Hospital under the intermediate supervision of Edmund G. Lowrie, M.D.

Permeability coefficients of 0.01 to 0.015 cm/min and ultrafiltration (adjustable over a range which included any clinical requirement) were demonstrated. These *in vivo* results were quite comparable to *in vitro* measurements.

The units themselves (both fibers and headers) were shown to be nonthrombogenic, although in one unit clotting may have been initiated at the onset of the procedure prior to the time blood reached the dialyzer.

The permeability rates for these fibers were comparable to published data on regenerated cellulose fibers.[11] However, total urea clearance was much lower than existing hollow fiber kidneys due to the lack of sufficient dialyzing area. Ultrafiltration rates for the modified PAN were many times higher at the same transmembrane pressures when data were compared on a per unit area basis.

Unit SBC-2. The patient being treated suffered from chronic renal failure. The initial hematocrit was 18.5 volume percent and 60 mg of heparin was administered intravenously at the start of the procedure.

The theoretical pressure drop/blood flow ratio for the device was 0.381 mm Hg/ml/min, assuming a blood viscosity of 2.5 centipoise. The unit was used as a dialyzer for 230 minutes. During the total time of the test, no significant deviation from the theoretical pressure drop/blood flow ratio was observed. It was therefore assumed that little or no fiber occlusion occurred during the course of the test.

At the conclusion of the test procedure, the inflow and outflow headers were examined with the aid of a dissecting microscope. There was a small amount of fibrin exuding from two or three capillaries in the outflow header. Several capillaries in the inflow header were occluded by thrombus but there was no evidence of clotting within the headers themselves.

A most significant observation was made which helped to amplify the excellent nonthrombogenicity of the hollow fibers themselves. It was observed that some of the fibers were originally occluded at a midpoint along their length in such a manner that blood was able to flow into the capillary but not through it. The blood pulsated back and forth in the capillary near this point of occlusion for the entire four hours of the *in vivo* test. Even though such a flow represented a significant stagnation point, clotting did not occur.

Mass transfer coefficients were estimated from the log mean mass transfer equation and are shown in Table 5. The mass transfer resistance versus the reciprocal of the dialyzate flow rate is also plotted in Figure 8. At relatively low dialyzate flow rates the mass transfer resistance was very high. On the other hand, as the dialyzate flow was increased, the resistance tended to fall. It would appear that the ordinate intercept of the curve in Figure 8 would be between 70 and 80 min/cm.

Results suggested, as was to be expected in the prototype model, that the efficiency of the dialyzate delivery could be improved to effect reduction in overall mass transfer resistance at lower dialyzate flow rates. However, at flow rates within the practical range, urea resistance was approximately 100 min/cm.

Ultrafiltration rates were estimated by observing patient weight loss with a Potter bed scale over given time intervals (a crude estimate). A plot of the transmembrane pressure versus ultrafiltration rate, shown in Figure 9, indicated that fluid losses from the small dialyzer were quite high. If the dialyzate inlet and outlet ports were redesigned, high ultrafiltration rates could readily be obtained, but would not be an obligation of the system if it was not clinically required.

Unit SBC-4. The test was carried out with single-pass dialyzate flow and a blood pump. The kidney was used initially in a horizontal position, then at a 45° angle (both non-recommended positions), and lastly in a nearly vertical position. There were no leaks, and patient clotting time was always greater than 30 minutes. However the initial blood reaching the dialyzer was not anticoagulated as heparin was administered into the dialysis tubing immediately after the start of the procedure rather than to the patient prior to the onset of the procedure. Clumps of red cells (observed using a hand lens) were noticed entering the dialyzer at the onset of the test.

The usual parameters of pressure flow and mass transfer were evaluated. In addition, white blood count, platelet count, serum haptoglobin and alpha I anti-trypsin were measured across the device.

Table 5

OVERALL MASS TRANSFER RESISTANCE FOR UREA AND CREATININE AND $\Delta P/Q$ WITH
TIME AND DIALYSATE FLOW FOR THE MRC PROTOTYPE HOLLOW FIBER DIALYZER*

Time (min)	30	55	80	110	155	180	210	225
Blood flow (ml/min)	90	92	89	89	92	90	89	90
$\Delta P/Q$ mm Hg/(ml/min)	.368	.362	.376	.352	.352	.337	.332	.337
R urea (min/cm)	156	109	104	95	89	94	95	93
R creat. (min/cm)	185	161	204	137	133	128	123	112
Dialyzate flow	480	670	670	930	1060	1060	1250	1250

*Tests conducted at Peter Bent Brigham Hospital

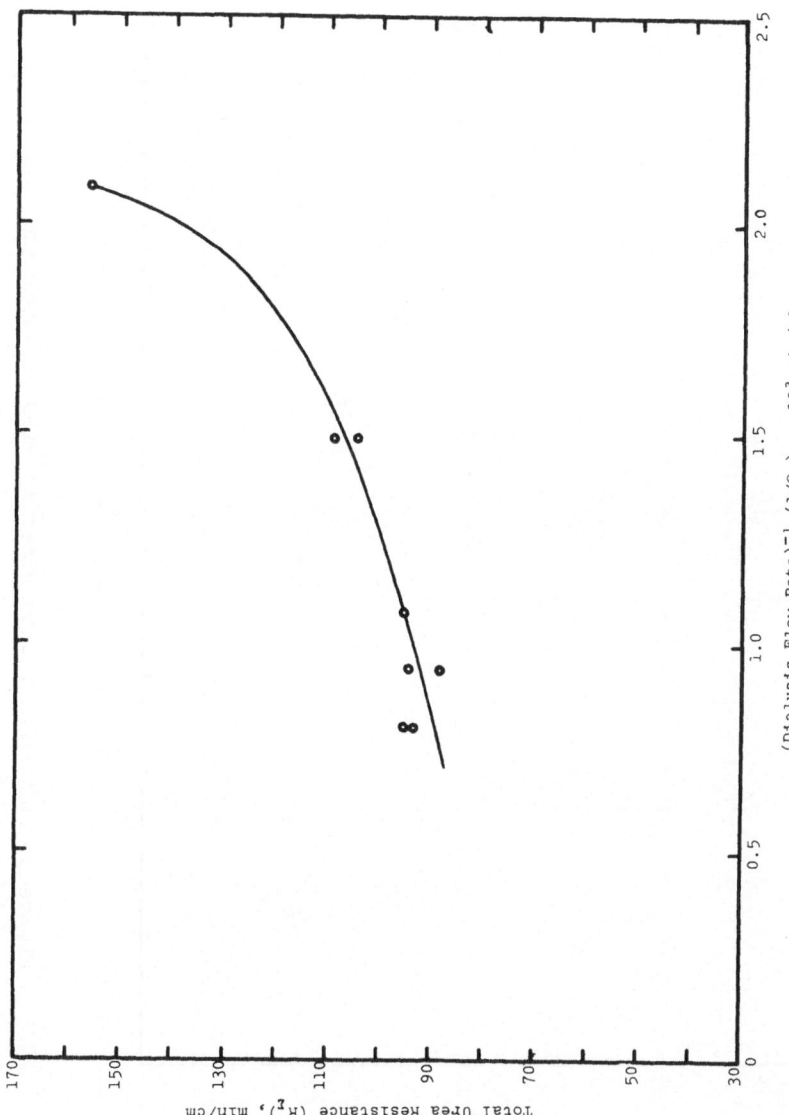

Figure 8.   Total Urea Mass Transfer Resistance as a Function of Reciprocal
            Dialyzate Flow Rate for Prototype SBC-2.

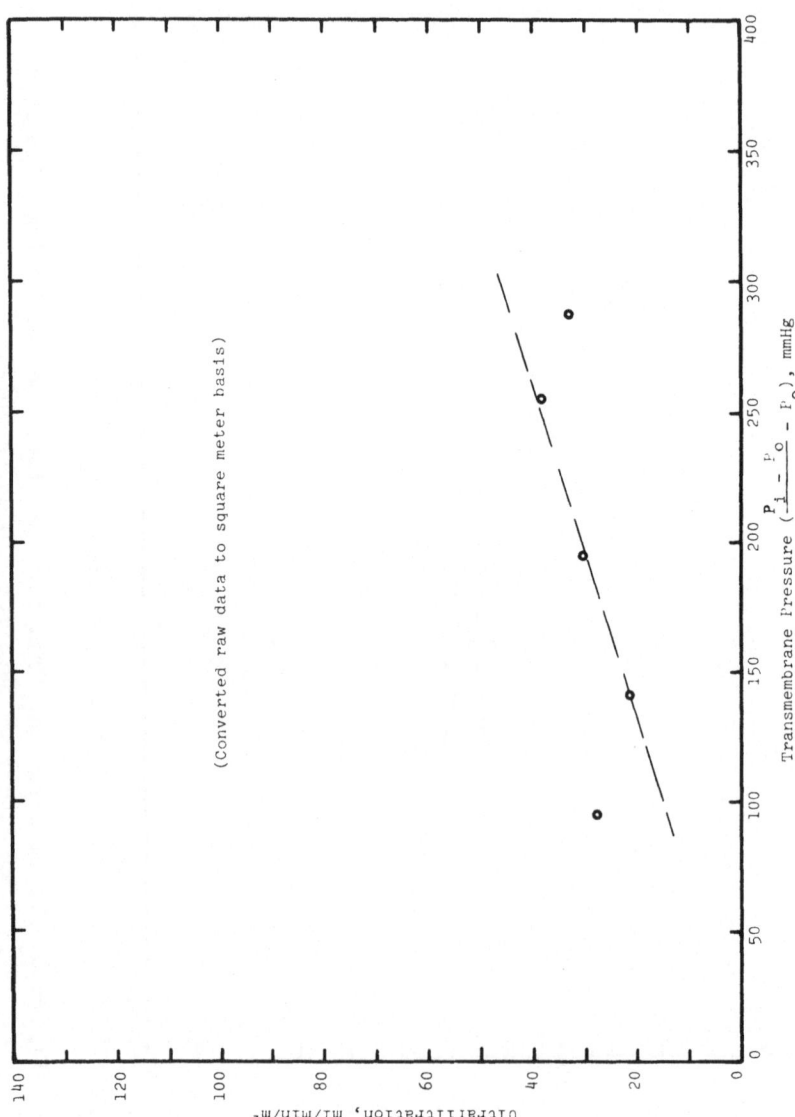

Figure 9. Ultrafiltration Rates *In Vivo* on Prototype SBC-2.

An initial pressure flow ratio of 0.506 mm Hg/ml flow (theoretical 0.245 mm Hg/ml flow) was noted, which rapidly climbed to
1.37 mm Hg/ml of flow and stabilized. The time difference between
these two observations was approximately 20 minutes. After approximately 60 minutes the unit was changed from the horizontal position
to a 45° angle and the reduction in the pressure flow to approximately 0.9 mm Hg/ml flow was noted. However, within the next ten
minutes this value increased to 1.2 mm Hg/ml flow. Even though the
device was eventually changed to the vertical position, it remained
in this range for the duration of the 180-minute test.

From a clotting standpoint, SBC-4 did not perform as well as
SBC-2. Fibrin was noted to be exuding from the capillaries in the
outlet header. This was no doubt due to the lack of initial heparinization of the patient, and the horizontal position of the
initial dialysis.

With the loss of surface area due to partial clotting, urea
and creatinine resistances tended to increase with time. Mass
transfer coefficients were estimated from the log mean mass transfer
equation making corrections due to the reduced surface, according
to the increase in pressure flow ratio. The value for urea was
approximately 70 min/cm, which was in accord with that observed
with the SBC-2 unit. Once again dialysis resistance contributed to
overall mass transfer resistance as was pointed out with prototype
SBC-2. Ultrafiltration measurements were not made during this run.

No significant differences across the dialyzer in white blood
cell count, platelet count, or serum haptoglobin was noted. Alpha
I anti-trypsin was not significantly affected. This suggested that
the blood elements were not adversely affected in their transit
across the device.

ACKNOWLEDGEMENTS

The work reported herein was sponsored by NIH-NIAMD Contract
No. PH43-67-1161 from June 1967 through June 1969.

We wish to acknowledge the direct technical contributions of
W. E. Weesner, D. D. Bump, L. E. Erbaugh, R. L. Leonard and J. S.
Tapp of Monsanto Research Corporation and the essential role of
J. P. Merrill, C. L. Hampers, and E. G. Lowrie, at Peter Bent
Brigham Hospital, as clinical evaluators.

REFERENCES

1. Stewart, R.D., et al., *Trans. Amer. Soc. Artif. Int. Organs*, 14, 121 (1968).

2. Stewart, R. D., et al., *Invest. Urol.* 3, 614 (1966).

3. Gotch, F., et al., *Trans. Amer. Soc. Artif. Int. Organs*, 15, 87 (1969).

4. Salyer, I. O., et al., Research of Materials and Devices for Artificial Kidneys, Annual Technical Progress Report, Monsanto Research Corporation, Contract No. PH-43-67-1161, National Institutes of Health, National Institute of Arthritis and Metabolic Diseases, June 1967 - June 1968, page 23.

5. Salyer, I. O., et al., Research on Materials and Devices for Artificial Kidney, Second Annual Technical Progress Report, Monsanto Research Corporation, Contract No. PH-43-67-1161, National Institutes of Health, National Institute of Arthritis and Metabolic Diseases, June 1968 - June 1969, page 29.

6. op cit., Ref. 4, page 34.

7. op cit., Ref. 5, page 31.

8. Salyer, I. O., et al., Materials and Components for an Artificial Heart, Annual Technical Progress Report, Monsanto Research Corporation, Contract No. PH -43-66-975, National Institutes of Health Artificial Heart Program, National Heart Institute June 1966 - June 1967.

9. Salyer, I. O., et al., Materials and Components for Circulatory Assist Devices, Second Annual Technical Progress Report, Monsanto Research Corporation, Contract No. PH-43-66-975, National Institutes of Health National Heart Institute, June 1967 - June 1968.

10. Salyer, I. O., et al., Materials and Components for Circulatory Assist Devices, Third Annual Technical Progress Report, Monsanto Research Corporation, Contract No. PH-43-66-975, National Institute of Health, National Heart Institute, June 1968 - October 1969.

11. Holmes, C. W., et al., Research on Hollow Fiber Dialyzer Units, Annual Summary Report, Dow Chemical Company, Contract No. PH-43-66-549, National Institute of Arthritis and Metabolic Diseases, National Institutes of Health, June 1967.

MEMBRANE REQUIREMENTS IN THE PURSUIT OF BLOOD OXYGENATOR OPTIMIZATION

Wilfred F. Mathewson, Ph.D. & David M. Ryon

General Electric Company

Medical Development Operation

Until recently, clinical use of blood oxygenators has been restricted to direct contacting devices, in which the blood has a direct interface with oxygen. This is most commonly accomplished by either passing bubbles of oxygen through a continuous phase of blood or filming the blood and allowing oxygen to contact the free surface. In both cases, addition of oxygen and removal of carbon dioxide from the blood is readily achieved. However, the high surface energies present at the gas/blood interface causes excessive blood trauma in the form of hemolysis and protein denaturation.[1] This limits such devices to short term heart/lung bypass such as open heart surgery where only several hours of perfusion are required. If clinical applications are to be enlarged to include intermediate and long term support for coronary diseases and pulmonary insufficiency, then means must be found to reduce blood trauma to acceptable physiological levels. Hopefully, this will be the role of the membrane lung.

The purpose of the membrane in a blood oxygenator is to reduce the energy at the free blood surface by placing a gas permeable film between the gas and blood phases.[2] Therefore, the membrane must not only be highly permeable to oxygen and carbon dioxide, it also must be highly compatible with blood. Materials used for this purpose to date have been porous Teflon and silicone rubber. More recently, block copolymers of silicone and polycarbonate* have proven suitable for this application.[3,4] Although gas transfer and blood trauma are of prime importance in the design of a membrane oxygenator, other clinical benefits can be derived from a properly designed unit.

_____
*General Electric MEM 213

59

By its very nature, a direct contacting oxygenator is an open system and it is difficult to control the relative volumes of blood in the patient and in the bypass circuit. Since the membrane lung is a closed system, this problem is eliminated. Furthermore, closure of the oxygenator system allows for a reduction in priming volume. This is of considerable importance where non cellular fluids are employed for prime since hemodilution is significantly reduced. For example, it is possible to design a membrane lung for infant application which will have a prime of 100 ml as opposed to an open system, direct contacting devices, which typically require from 500 to 1000 ml .... nearly the blood volume of the infant alone. Therefore, the combination of low blood volume and blood volume control is of obvious benefit to the clinician.

Before the design of a membrane lung can be pursued in more detail, it is necessary to focus on the human lung function and its performance characteristics. The hemoglobin oxygen saturation increase across the lung can be approximated by the following:

$$\%\Delta AV = \frac{22.5 \text{ X (expected } O_2 \text{ transfer } - \text{ ml/min)}}{(\% \text{ hematocrit}) \text{ X (blood flow rate } - \text{ liters/min)}} \qquad (1)$$

For an 80 Kg adult, the basal O_2 consumption is 250 ml/min and the cardiac output is 5.6 liters/min. Therefore, with a normal hematocrit of 40%, the A-V difference across the lung is 25%. For an infant of 10 Kg, basal O_2 consumption is 75 ml/min and the cardiac output is 1.7 liters/min. Here again, the A-V difference is 25%. Table I summarizes these oxygen requirements for the membrane lung. All of these requirements must be met if the membrane lung is to be of clinical significance.

TABLE I

Oxygen Transfer Design Requirements

Oxygen Transfer - up to 250 ml/min

% O_2 Saturation Difference - greater than 20%

Blood Flow Rate - up to 6 liters/min

Oxygen Transfer To Blood Flow Ratio - 40 to 50 ml O_2/liter blood

For partial support, the commensurate CO_2 requirements can not be determined a priori for reasons brought forth in Figure 1.[5] The human lung is overdesigned in O_2 transfer as illustrated by the O_2 gradient in an alveolar capillary. However, on inspection of the CO_2 gradient in the capillary, it is obvious that the lung is significantly more overdesigned in CO_2 transfer capability. Therefore, in partial support applications where lung function is partially lost, it is most likely that O_2 transfer will be limiting long before CO_2 transfer. This circumstance places a minimum demand on the membrane lung with regard to CO_2 requirements. However, for practical clinical considerations, membrane lung design must include the possibility of total lung support. In such a case, CO_2 must be removed from the blood at a rate approximately 80 to 100% of the rate of O_2 transfer, depending on what foods the body is metabolizing, i.e. carbohydrates, fats, or proteins. That is, the respiratory quotient falls between 0.8 and 1.0 depending on the body's composite metabolic process. Therefore, an additional membrane lung requirement must be added to those in Table I to cover this physiological requirement.

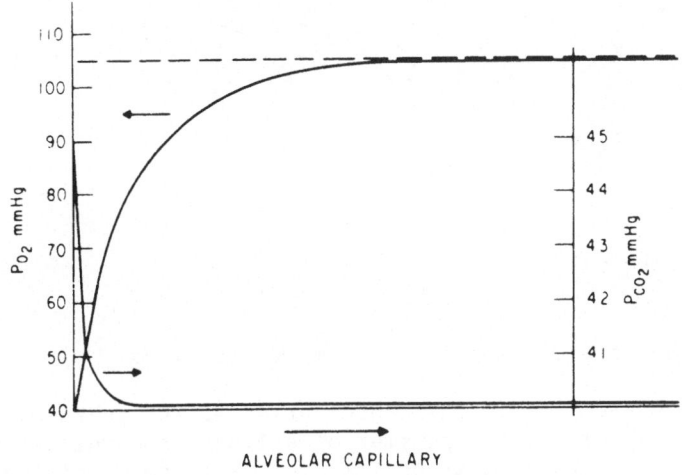

FIGURE 1

Gas Gradients In The Alveolar Capillary

In order to gain insight into design considerations affecting gas exchange, it will be useful to inspect the design and performance of a planar configuration - the General Electric Peirce lung. A schematic of the lung design is shown in Figure 2. Each pair of membrane surfaces is combined with two distribution discs to form a membrane envelope, all of which is supported with an external vinyl frame. The distribution discs are located at diagonal corners of the envelope, and when envelopes are stacked to form a package, these discs become an integral part of the primary blood manifold. Oxygen distribution to each side of the membrane envelope is provided by the use of a composite support consisting of a woven plastic screen covered with a thin protective layer of non-woven material. This support is shaped such that there is a taper between the support and the outer frame. This gap allows for the distension of the membrane in this unsupported region, and thus forms a tapered channel for uniform flow of blood from the distribution disc to the transfer area of each membrane envelope.

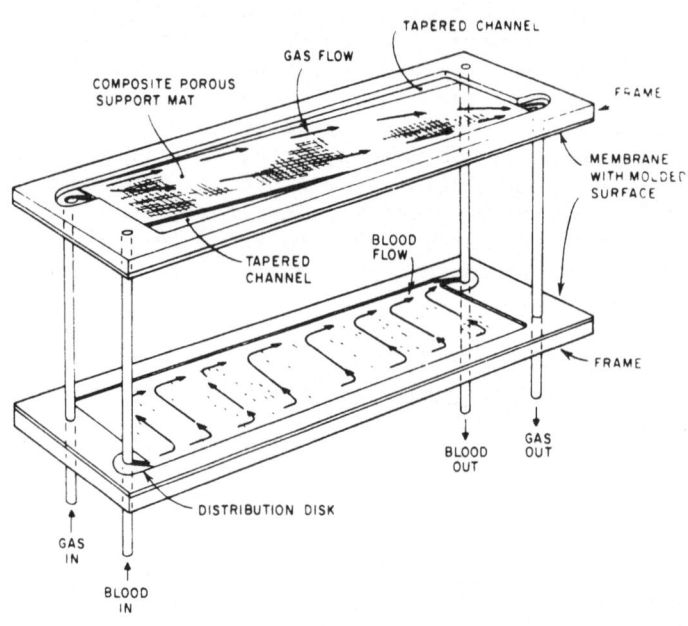

FIGURE 2

Schematic Of General Electric Peirce Lung

The membrane employed in the construction of the package is
MEM 213: a General Electric copolymer of silicone and polycarbonate.
This material was chosen because of its excellent compatibility with
blood.[3,6] All present indications are that MEM 213 is equal to or
better than dimethyl silicone with regard to hemolysis, clotting
factors, and protein denaturation. Furthermore, MEM 213 being a
thermoplastic provides the designer with a wide range of processing
alternatives. In the case of this package design, this processing
latitude has been employed to produce a molded membrane surface.
This surface provides for positive control of the separation of the
membranes over the entire active area; which, in turn, insures uniform
blood flow. Furthermore, this surface provides for gentle mixing and
secondary flow of blood to enhance oxygen transfer.

All of the experimental results obtained on this package design
have been carried out employing an homologous lung procedure.[7] As
can be seen from inspection of Figure 3, two pumps are used in the
circuit. One of these pumps is employed as a recirculating pump,
while the other is the main arterial pump. In practice, the venous
pump is adjusted to provide flow to the lung at a rate slightly
greater than the maximum anticipated. During the procedure, the
arterial pump is then adjusted to the desired flow rate.

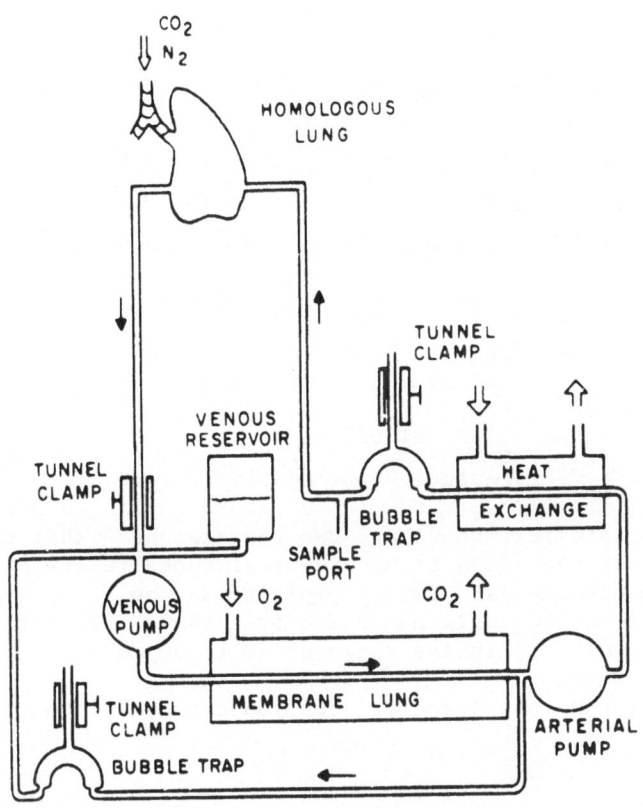

FIGURE 3

Schematic Of Blood Flow Circuit

The oxygen saturation increase across the membrane lung was measured directly with the use of an American Optical reflecto-meter. The results are shown in Figure 4. The four models tested were PLM 100, 500, 1000, and 2000 designed for blood flow rates respectively of 1/10, 1/2, 1, and 2 liters/min. The oxygen trans-fer rate was calculated from equation (1) from measured values of A-V, blood hematocrit, and flow rate. The results are shown in Figure 5; again, for the four models tested.

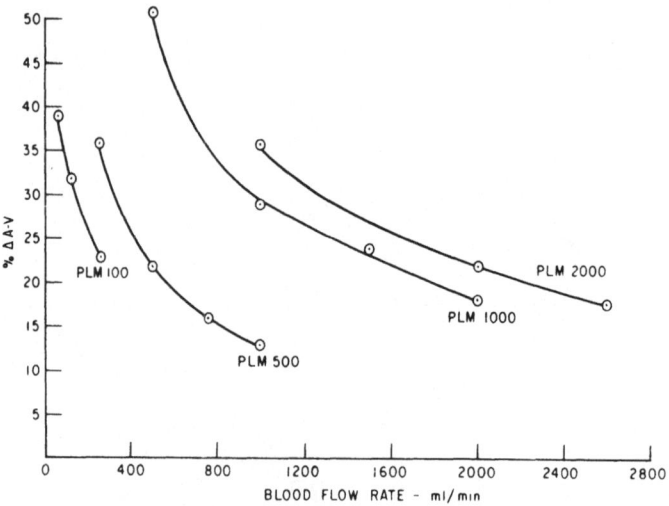

FIGURE 4

A-V Difference Vs Blood Flow Rate

In oxygen transfer, there are two resistances - that of the membrane and that of the blood film. The resistance in the gas phase is nil because pure O_2 is normally employed as the sweep gas and the partial pressure of CO_2 will never increase above its equilibrium value in blood which is in the range of 20 to 50 mm Hg.

Therefore, the mass transfer equation can be written as:

$$N_{O_2} = K_{O_2} A (P_g - P_b)_{O_2} \qquad (2)$$

where N_{O_2} is the oxygen transfer rate - cm^3/sec, K_{O_2} is the overall transfer coefficient for the membrane and blood film - cm/sec, atm, A is the active area of mass transfer - cm^2, and $(P_g - P_b)_{O_2}$ is the difference between the partial pressure of O_2 in the gas phase and the partial pressure of O_2 at equilibrium with the bulk of blood - atm.

The overall coefficient can be related to the individual transfer coefficients by:

$$\frac{1}{K_{O_2}} = \frac{tm}{P_{r_{O_2}}} + \frac{1}{\alpha_{O_2} k_{O_2}} \qquad (3)$$

where $P_{r_{O_2}}$ is the O_2 membrane permeability - cm^2/sec, atm, tm is the membrane thickness - cm, α_{O_2} is the solubility coefficient of O_2 in blood plasma - $cm^3 O_2/cm^3$ plasma, atm, and k_{O_2} is the blood mass transfer coefficient - cm/sec.

As would be expected in viscous flow, O_2 transfer is dependent on flow or residence time. For any given design geometry, the blood mass transfer coefficient will be approximately inversely proportional to the contact time to the 1/3 power.[8] Therefore:

$$k_{O_2} = \frac{CD_v}{\theta^{1/3}} \qquad (4)$$

where C is a design constant, D_v the diffusion coefficient of O_2 in plasma — cm^2/sec and θ the contact time — sec.

From the oxygen transfer data shown in Figure 5, it is possible to calculate K_{O_2} as a function of flow. The appropriate contact time can be calculated from the measured initial blood volume divided by the flow rate. From a combination of equations (3) and (4), it can be seen that if the reciprocal of K_{O_2} is plotted against $\theta^{1/3}$, the intercept will be the membrane resistance and the slope will be linear and equal to $1/\alpha CM_v$. The result of the data treated in this manner is shown in Figure 6. The time base for this plot was chosen as minutes rather than seconds to simplify presentation. The intercept of 70 min, atm/cm comes reasonably close to the theoretical membrane resistance of 78 min, atm/cm for 1.5 mil MEM 213. The results not only show a good correlation to theory, they also show that there is little size factor associated with this particular planar design since the units investigated covered lung packages consisting of from four to sixteen membrane envelopes. Furthermore, the results also demonstrate that over the blood flow rates of interest for each model, the blood phase resistance dominates — being 75 to 80% of the total resistance.

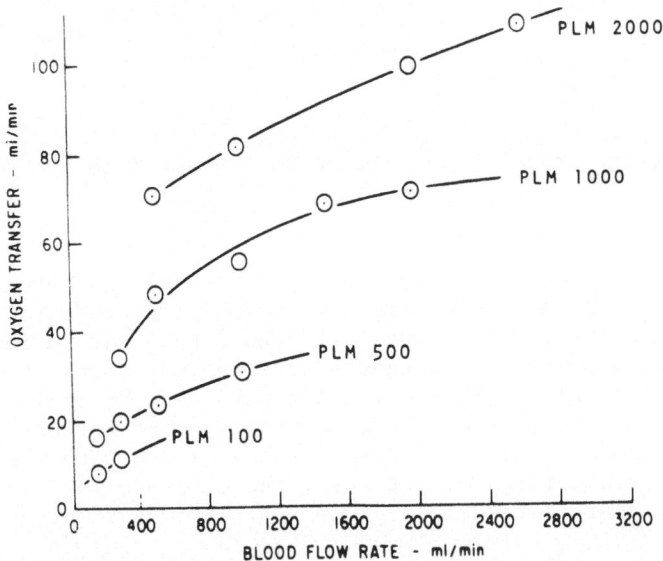

FIGURE 5:   Oxygen Transfer Vs Blood Flow Rate

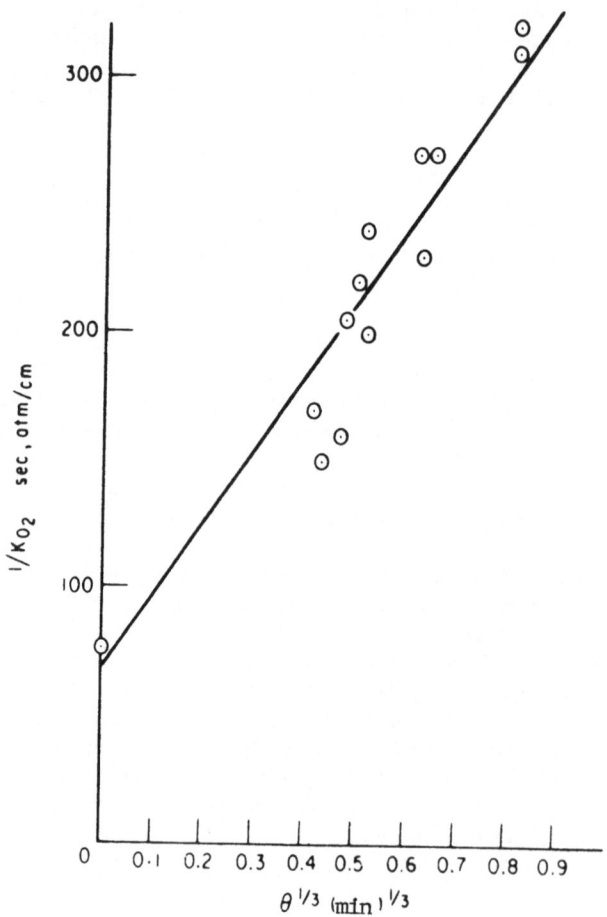

FIGURE 6

Oxygen Transfer Resistance Vs Contact Time

The CO_2 transfer measurements were not nearly as accurate as those for O_2 since they were estimated from a Singer-Hastings nomogram [9] from measured pH values corrected for O_2 saturation. The average of all measurements suggests that the CO_2 transfer rate is 80% of the O_2 transfer. This being the case, and by employing equations (2) and (3), it can be demonstrated that for CO_2 removal, neither the membrane nor blood film resistance dominate - since in the blood flow rate range of interest, the membrane resistance is 55% of the total resistance.

A greater perspective can be gained as to the relative importance of membrane and blood film resistance by inspection of the ratio of CO_2 to O_2 transfer over a range of blood film mass transfer coefficients. In order to achieve this, equations (2) and (3) are written for both O_2 and CO_2 and combined to yield:

$$\frac{N_{CO_2}}{N_{O_2}} = \frac{(P_b - P_g)_{CO_2}}{(P_g - P_b)_{O_2}} \times \left[\frac{\dfrac{tm}{P_{r_{O_2}}} \times \dfrac{1}{\alpha_{O_2} k_{O_2}}}{\dfrac{tm}{P_{r_{CO_2}}} \times \dfrac{1}{\alpha_{CO_2} k_{CO_2}}} \right] \qquad (5)$$

where:

$$(P_b - P_g)_{CO_2} \simeq 0.053 \text{ atm}$$

$$(P_g - P_b)_{O_2} \simeq 0.92 \text{ atm}$$

$$P_{r_{O_2}} = 12 \times 10^{-7} \text{ cm}^2/\text{sec, atm}$$

$$P_{r_{CO_2}} = 74 \times 10^{-7} \text{ cm}^2/\text{sec, atm}$$

$$tm = .0025 \text{ \& } .005 \text{ cm}$$

$$\alpha_{O_2} = 0.026 \text{ cm}^3 O_2/\text{cm}^3, \text{ atm}$$

$$\alpha_{CO_2} = 0.54 \text{ cm}^3 CO_2/\text{cm}^3, \text{ atm}$$

Therefore, the only remaining independent variables are the mass transfer coefficients of O_2 and CO_2 in blood. If these values are assumed to be approximately equal over the conditions of interest, the respiratory quotient of the membrane lung will be solely dependent on the blood film coefficient for any given membrane material and thickness.

From the previous analysis, it is possible to locate the average value of gas exchange experiments with reference to the CO_2/O_2 ratio and the blood film coefficient in the range of design flows for the various size units. As shown in Figure 7, this averaged value lies close to that expected theoretically for a 1.5 mil film of MEM 213.

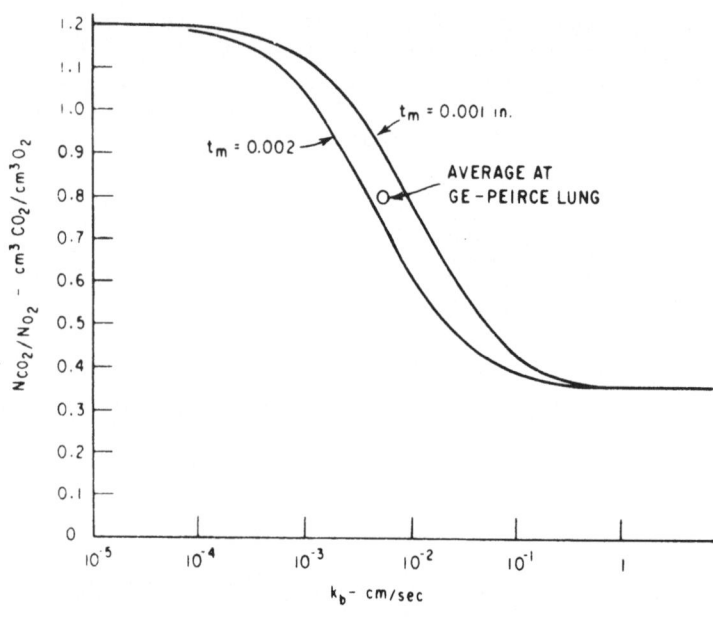

FIGURE 7

Gas Transfer Ratio Vs Blood Transfer Coefficient

The analysis also demonstrates that when gas transfer is blood film limiting, the CO_2/O_2 ratio will have a maximum value of 1.2. The minimum ratio of .35 is reached when the membrane limits the exchange process. From practical clinical reasons mentioned earlier, it is important to keep the CO_2/O_2 ratio above 0.8. Therefore, if the performance of the planar membrane lung is to be improved, it is necessary to affect two changes simultaneously; decrease the membrane thickness and increase the blood transfer coefficient. The former will no doubt be achievable as we learn more of the processing technique to fabricate pin-hole-free polymer films at the required thicknesses of 1 mil and less. In order to improve the blood coefficient and still maintain the requirements outlined in Table I, it will be necessary to reduce the blood channel height. At the present time, for the planar design in question, the blood film is slightly less than 5 mils. If, as required for higher gas transfer performance, the thickness is reduced, it will be necessary to consider the engineering and practical effects this alteration will have on pressure drop across the lung.

The pressure drop-flow characteristics for the present design are shown in Figure 8. It will be noticed that at design flows, the pressure drop is around 120 mm Hg. A portion of this loss occurs thru viscous losses as blood flows across the active membrane surface. However, of equal importance and necessity, there

are also viscous and momentum losses required to properly manifold
the units such that every membrane surface has the same characteristic flow residence time. The proper balance of these two loss
factors is of prime importance if the design is to achieve efficient membrane utilization. When one considers that the viscous
losses across the membrane surface are proportional to the blood
film thickness to the third power, it becomes obvious that pressure drop consideration will dictate much of the redesign logic
necessary to achieve higher O_2 and CO_2 transfer rates.

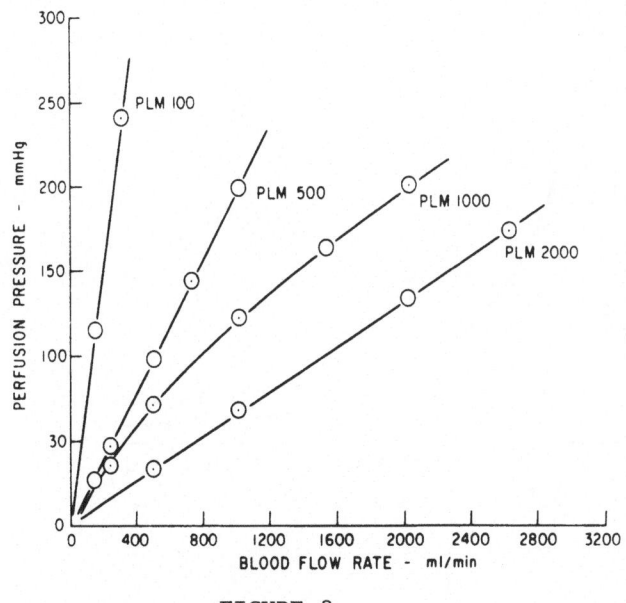

FIGURE 8

Perfusion Pressure Vs Blood Flow Rate

An alternative design configuration for a membrane lung is a
capillary gas exchanger in which blood flows thru membrane capillaries while oxygen is passed over and around the capillaries.
This system has been investigated by Weissman & Mockros[10] and
they have found that membrane resistance to oxygen transfer is
negligible for reasonable values of membrane wall thicknesses.
This is true because the ratio of membrane area to blood volume is
higher for a capillary configuration than that of the planar design
discussed in this paper. However, in order to obtain oxygen transfer rates in the range of those obtained with the General Electric
Peirce lung at equivalent A-V differences, it is necessary to
have capillaries with an inside diameter of 8 mils or less. This
is shown in Figure 9 where the oxygen flux is plotted against capillary diameter for A-V differences of 20 and 25%.

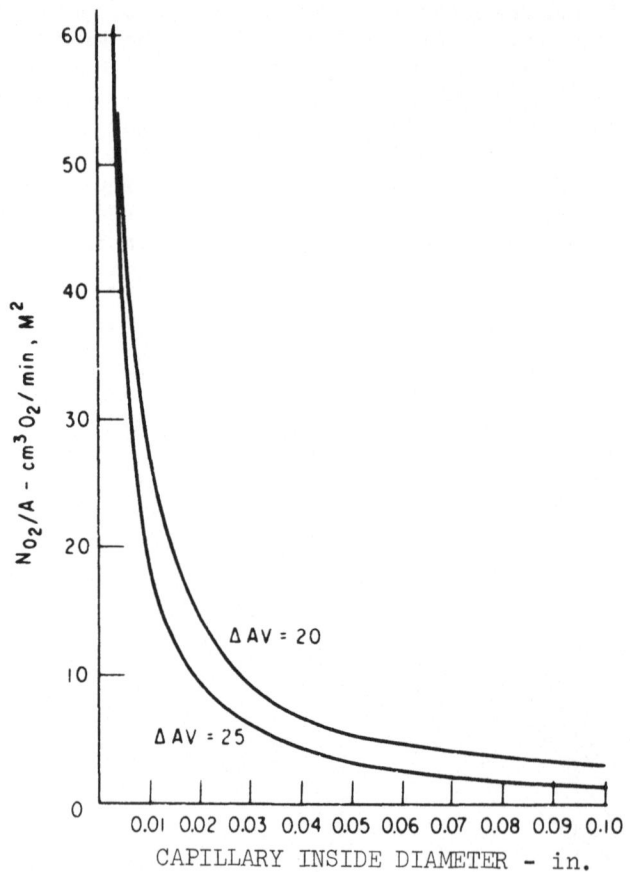

FIGURE 9

Oxygen Flux Vs Capillary Diameter

Those interested in a capillary oxygenator, then, are confronted
with several unique problems in the design and fabrication of such
a device. First, there is the obvious problem of extruding capillaries in the sizes required with materials like silicone rubber
and silicone polycarbonate copolymer at economic rates. Furthermore, the designer must provide an adequate means of manifolding
thousands of capillaries without inducing preferential flow and
without generating a significant clotting factor at or near the
capillary entrances. Based on the evidence to date of planar vs
capillary performance, it is questionable whether an attack of the
practical problems associated with capillary designs is warranted.

Today there are several models of planar membrane blood oxygenators available to the clinician for his evaluation. Some of

these models are capable of meeting the physiological oxygen and
carbon dioxide transfer requirements sufficiently to warrant their
use in on-going open heart procedures. Expansion of heart/lung
bypass beyond open heart surgery is not dependent on further
improvement of gas exchange capability of planar or capillary
designs. Rather, expansion is dependent on the development of
suitable clinical protocols by clinical teams who are keenly interested in the potential benefits of long term heart/lung support.
In the main, the direction of future development of the membrane
lung should come through feedback from these clinical teams.

REFERENCES

1   Gollub, S. and Bailey, C.P., JAMA, 198, p. 1171, (1966).

2   Peirce II, E.C., "The Membrane Lung - Its Excuse, Present
     Status and Promise", Journal of the Mount Sinai Hospital, 34,
     No. 5, (Sept. 1967).

3   Peirce II, E.C., et al, Trans. Am. Soc. Artificial Internal
     Organs, 15, pp. 33-38, (1969).

4   Peirce II, E.C., Trans. Am. Soc. Artificial Internal Organs,
     16, (1970).

5   Guyton, A.G., Textbook of Medical Physiology, pp. 574-578,
     W.F. Saunders, Philadelphia, (1968).

6   Mason, R.G., et al, National Institutes of Health - Artificial
     Heart Program Conference Proceedings, pp. 193-202, (1969).

7   Peirce, E.C. and Mathewson, W.F., National Institutes of
     Health - Artificial Heart Program Conference Proceedings,
     pp. 405-416, (1969).

8   Bird, R.B., et al, Transport Phenomena, pp. 349-370, John
     Wiley & Son, New York, (1960).

9   Hastings, A.B. and Singer R.B., Medicine, 27, No. 2,
     pp. 223-243, (May 1948).

10   Weissman, M.H. and Mockros, L.F., Med. & Bio. Eng., 7,
     pp. 169-184, (1969).

MEMBRANES FOR PRESSURE PERMEATION

Henry Z. Friedlander and L. M. Litz

Union Carbide Research Institute,

Tarrytown, N. Y. 10591

I. Introduction

Molecular separations of gases by passage through a thin, dense membrane have been known for at least 140 years.[1] Even the desalination of water by passing it through a membrane under pressure with the exclusion of dissolved salts was known and named reverse osmosis 45 years ago.[2] Yet it was not until the 1960's that the latter became an important, practical process. For one thing, identification of water supply as a vital problem requiring public support for alternative sources was necessary. Following this awareness came seven-digit investment in laboratory research by public agencies such as the State of California and the U. S. Department of Interior which led among other advances to the asymmetric membrane.[3,3a] This development cut the area of membrane surface required and hence the cost of an installation for a given volume processed to one-hundredth of that previously thought necessary. With the arrival of submicron, thin membranes a new unit operation was born.[4,5,6,7]

In selecting materials for membranes many physical, chemical, and economic criteria must be kept in mind. Three of these are fundamental and remain the cornerstone in the thinking of the research worker considering membranes for a given separation. They are flux, semipermeability, and mechanical strength.

This paper considers molecular separations made by semipermeable membranes in which a pressure gradient across the membrane is the driving force for the process. This operation is usually termed pressure permeation, or reverse osmosis in the case of aqueous separations. When the downstream face of the membrane is

subjected to a partial vacuum, the process is called pervaporation
(permeation + evaporation). Since the physical form and macromolecular morphology of such membranes are inextricably related to
the design of the membrane element and the permeator, one must constantly be aware of the physical situation while considering the
choice of membrane material.

Although the separation of large molecules, colloids, or
particles utilizing comparatively open diffusive membranes can be
operated at low pressures of about 25 to 150 psi (ultrafiltration),
molecular separations require tighter membranes and, hence,
typically a pressure of 300-2000 psi in order to get acceptable
permeability. At these pressures almost all membranes must be supported in either a flat plate, tubular, bilateral septum,[4] or
spiral-wound modular configuration.[5] The only type of design
which does not require an additional support is the hollow fiber
membrane which acts as its own support.[6]

The membrane used in these designs can itself take one of
several forms, namely; sheet, coating, tubular, hollow fiber,
colloidal, or liquid. Generally the membranes used today are organic, high-molecular weight polymers, but glass, metals, monomeric liquids or various particulate materials can be used. The
tubular, hollow fiber, and liquid membranes are limited in application to a single physical form. Colloidal membranes have been
made in flat plate as well as tubular devices, while polymer sheets
and coatings have the most versatility.

Among the various forms of membranes both the flat sheet and
tubular versions can be made with an asymmetric profile. An asymmetric (graded, skinned, vectored, anisotropic) membrane consists
of a thin, dense layer integral with a thick, supporting spongy
layer. The thin layer is often about one-third of a micron thick
while the supporting layer is usually 75-100 microns thick. This
tailored profile is formed at the time of casting from a complex
solution of the polymer. At this writing asymmetric membranes of
cellulose diacetate, cellulose acetate butyrate, polyacrylonitrile,
poly(vinyl carbonate), and an aromatic polycarbonate are the only
known ones to the public.

A similar result can be achieved by making separately a submicron, thin film and placing it on a separate support as a lami-
nate. Some authors term this a "composite" membrane, although that
phrase might also appear to apply to membrane materials, dip-coated,
printed, or electrostatically sprayed onto porous supports.

This last concept leads us toward the process of forcing a
wide variety of colloidal materials onto or into microporous supports so that these colloids act like films even though particulate
or colloidal in nature. These non-film, colloidal (dynamic, sedi-

mentary) membranes can make separations at the molecular level. When the material being separated from the liquid medium is itself the colloidal membrane, as with sewage or sulfite paper wastes, then it may be called an autogenic colloidal membrane.

Finally, one should mention various surface active or polymeric additives which can be added in minute amounts to the liquid being processed in order to improve membrane performance. These high molecular weight or polymeric additives may function as a colloidal membrane superimposed on a normal membrane film (usually asymmetric).

II. Membrane Materials

A. Desalination with Cellulosics

The most widely studied membrane separation process activated by pressure is desalination by the "reverse osmosis" process. [7,8,9,10] The key development in this area occurred ten years ago with the advent of the asymmetric cellulose diacetate membrane. The original desalination experiments employed duPont CA-43 cellulose acetate. [11] Later the most widely used material was Eastman's cellulose 2.46 acetate, CA-398. Recently there has been increasing interest in CA-394, slightly lower in acetyl content. At this writing the Eastman, Gulf-General, Universal Water, AMF, Aerojet, Havens, and American Standard organizations are believed to produce versions of this membrane in sheet or tubular form on a commercial scale. The importance of this development is underscored by the realization that although hundreds of polymers have been evaluated in tens of laboratories in the last decade, secondary cellulose acetate is still the polymer of choice.

Easily soluble in many solvents (e.g., acetone, DMF, dioxane, pyridine, nitromethane, acetic acid, methylene chloride, butanone, methyl acetate) cellulose diacetate has a high transition at about 207° (and lower ones at about 60° and about 115°), a stiff chain due to the rings in the polymeric backbone, a long segmental chain length, and extensive hydrogen bonding. It's water sorption of about 14-17 percent apparently is ideal for transporting water with the rejection of most ionic salts. This last statement, even if true, does not help in answering the basic question: exactly why is secondary cellulose acetate the best membrane material known today? The morphological aspects first mentioned must be stressed because, as will be detailed later, cellulose acetate is an excellent membrane material for separating organic and gaseous mixtures as well. The high transitions, partial crystallinity (opinions vary from 30-60 percent), and anisotropic structure resulting in a high free-volume come close to answering the basic question concerning the critical morphology for generalized semipermeability.

Cellulose diacetate does have several drawbacks, however.
The asymmetric structure is subject to compaction under the 600-
1500 psi pressure generally used. The rapid loss of flux at 1500
psi makes its use for direct sea water desalination unattractive.

Finally, cellulose acetate is subject to microbial attack as
well as hydrolysis at pH's much removed from neutrality.[12]   In
the last few years much study has gone into improving the original
Loeb-Sourirajan membrane. This is summarized in references 13, 14,
and 15 as well as the Saline Water Conversion Reports. Some of the
more important modifications are discussed below:

In order to reduce compaction one can crosslink the cellulose
acetate membrane with a chemical agent such as formaldehyde, a di-
isocyanate, a substituted melamine derivative (Cymel 300), a
titanium chelating agent (Tyzor TE), a sulfone (Sulfoset 60), or
by high energy radiation.[16]   The incorporation of inorganic
fillers or pigments may serve the same end. Also a crosslinked,
polymeric microgel filler may be employed.[17]   Another mode of
crosslinking is to substitute partially a polymerizable acyl moiety
in place of some of the acetate groups on the cellulosic backbone.
Vinyl polymerization of the substituent (e.g., methacrylate) im-
proves the physical properties of the membrane.[18]   These approaches represent reasonably successful attempts to lower physical
creep under pressure.

Another approach to conquering the compaction problem, which
also gives higher salt rejection, is to provide an asymmetric
member of higher degree of acetylation. This can be done by blend-
ing cellulose triacetate with the usual cellulose 2.46 acetate, or
by obtaining the requisite acetate homopolymer. Apparently cellulose 2.59 acetate is best from the standpoint of minimal flux
decline ascribed to compaction as well as high enough salt rejection, 99.8 percent, so as to desalinate sea water in one pass.[19]
Cellulose acetate butyrate, which can be asymmetrically cast[20]
and is adequately rejecting for sea water desalination, is said to
show diminished compaction.

Chemical modification of the asymmetric membrane or of the
original polymer by graft copolymerization with styrene represents
another means of improving flux stability by decreasing compaction,
polystyrene being a "hard", brittle polymer. With 25-30 percent
styrene content decreased compaction is achieved with some sacri-
fice of either flux or rejection.[21]

Another major modification of the original Loeb-Sourirajan
asymmetric casting approach to   divorce the submicron, dense,
semipermeable layer from the 75-100 micrometer support. Once this
is elected, the two components need not necessarily be chemically

the same. In this way salt rejection and flux stability can be optimized individually. One can take advantage of the higher salt
rejection of cellulose triacetate by coating a thin film of it on
either cellulose or nitrocellulose.[22] This same approach is
valid for the hollow fiber form of membranes as well as flat sheets
or tubes. Thin layers of cellulose acetate 500 to 4,000 Å thick
can be cast onto water, removed, and then laminated to a microporous support such as polysulfone.[23] A wide variety of cellulosic mixed esters, ethers, and ionic derivatives have been
studied with encouraging but not spectacular results.[24] A
variety of thin film materials including poly(acrylic acid), poly-
(vinyl pyrrolidone), cellulose diacetate, cellulose triacetate have
been laminated over a cellulose nitrate — cellulose acetate support
by a variety of techniques in another study[25] to make so-called
"composite" membranes.

     Admittedly cellulose acetate was originally identified as an
excellent desalinating membrane material by a strictly empirical
approach.[11] The only guidepost being "it was felt that the
desired membrane should contain some hydrophilic groups." Most of
the discussion in the past has been devoted to the water-binding
properties of cellulose acetate along with analyses of how this
polymer-water system neatly combines features of sieving, hydrogen
bonding, water sorption, and ion-exchange models of water transport
in order to achieve high salt rejection. Many other polymers have
about the same solubility parameter as cellulose acetate — 11.1;
some are epoxy resins, poly(ethylene terephthalate), polymethacry-
lonitrile, poly(vinylidene chloride), as well as ethyl cellulose
and cellulose nitrate. But these materials do not make particularly good desalinating membranes.

     Neither this nor any other simplistic approach suffices to
define in advance the chemical composition for a desalinating
polymer. The hydrophilicity aspect of the problem is illustrated
by some data shown in Table I on various degrees of substitution
of cyanoethyl cellulose.[6] Although the water flux is strongly
dependent on the equilibrium water content, which is apparently
set by the degree of substitution, the salt rejection is essentially independent of this factor. The morphology as well as the
chemistry must play an important part in defining the permselectivity of such cellulosics.

     B. Desalination with Non-Cellulosic Polymers

     Vinyl polymers of intermediate polarity, especially when cross-
linked, form the next best class of desalinating membranes. Cross-
linked acrylic and methacrylic esters have been studied by many
workers, (e.g., 26, 27, 28). The two classes of polymers are combined in copolymers of methyl methacrylate and galactose methacrylate.[29] Poly(acrylic acid) and poly(vinyl pyrrolidone) have

Table I

Desalination by Cyanoethyl Cellulose of Various Degrees
of Substitution

Degree of Substitution	Equilibrium Water, Percent	Flux at 1000 psi gf^2d-mil	Percent NaCl Rejection
1.9	16.0	0.079	95
2.1	11.4	0.074	97
2.4	8.0	0.038	97
2.5	6.7	0.02	94
2.7	4.7	0.003	90
3.0	4.5	0.015	94

already been mentioned. Poly(vinyl alcohol) has been studied for
over ten years,[11,30] as have various polyethers.[15,26,30] Poly-
(vinyl alcohol) is attractive for study because it is hydrolytical-
ly stable as well as having many reactive sites. Heating this
polymer to 150° to insolubilize it gives a material with salt rejection of about 25 percent, while crosslinking by reaction with
acetaldehyde can give rejection up to 95 percent. Heated syndio-
tactic polymer showed about 50 percent salt rejection. Copolymers
of PVA containing 12-28 percent ethylene also had intermediate rejections of 25-60 percent.[6]

     In addition to cellulose acetate and cellulose acetate buty-
rate, asymmetric membranes have been cast of polyacrylonitrile and
poly(vinylene carbonate). Various copolymers and hydrolysis products of these polymers have been synthesized.[31]

     Among the newer polymeric systems studied one might mention
sulfonated poly(2,6-dimethylphenylene oxide)[32] block copolymers
of the sulfonated styrene-isoprene-styrene type,[33] and membranes
containing graphitic oxide.[34]

     In concluding this selective survey of a few of the hundreds
of polymers studied for their desalinating properties one might
mention the proprietary nylon of unpublished composition incorporated in the hollow fiber system of the duPont Co. This membrane is
used in the 100,000 gallon per day plant started up in June, 1969
at Plains, Texas.[35]

III. Colloidal Membranes

     The essence of these arrays of discrete colloidal material
derives primarily from their physical form rather than their
chemical composition. When a wide variety of inorganic or organic
polymeric material in finely divided form is caught as a thin layer
in or on a microporous support, it acts like an unbroken film and

can make molecular separations. This class of membrane is characterized by high fluxes (up to 20 times that of film membranes) and
moderate to good rejections of salts, sugars, or dissolved color
bodies, in the 20-95 percent range of rejection.

Much of the published work in this field has come from the Oak
Ridge National Laboratory based on laboratory work started in June,
1965.[36-40] The original colloid studied by this group was boiled
zirconium oxychloride which can polymerize to form a colloidal hydrous zirconia.[41] Other materials successfully used to make such
"dynamic" membranes range from bentonite and pulverized ion-exchanged
resin to poly(acrylic acid) or starch sulfate. Microporous supports include carbon tubes, ceramic tubes, or even firehose.

Earlier work includes a patent for underground desalination
by particulate membranes in porous subsurface formations.[42] This
same group describe semipermeable particulate membrane-forming
materials 0.1 to 100 micrometers in size in two categories in
another patent.[43] One category of "osmotic materials" includes
clays, ion-exchange resins, starch, cellulose, sulfonated polystyrene, salts of polyacrylates, and lignosulfates. Another "bound
water" category includes cellulose acetate, cellulose acetate butyrate, poly(vinyl alcohol), poly(vinyl acetate), polystyrene, and
polycarbonates. The mechanism of desalination in the second case
is explained in terms of water losing its solution capability for
salts by being bound to the polymer.[44] Other early work was done on
colloidal membranes by P. Kollsman and by a group at Dorr-
Oliver Co. who, in 1964, while studying sewage treatment realized
that a thin layer of sewage components and not the underlying
synthetic polymer was the true separating membrane. Besides sewage
effluents, sulfite paper mill wastes can, with or without added
filter aids, make autogenic colloidal membranes.[39,45]

Implicit in the concept of colloidal (dynamic, sedimentary)
membranes are two hydrodynamic features. Firstly, the velocity of
the upstream feed passing by the microporous support limits the
thickness of the membrane to a submicron layer. Secondly, this type
of membrane is self-healing since any incipient defect is immediately
filled by excess colloidal material. Only small amounts of colloidal material, often as little as 1 mg/20 cm^2 (50 mg/ft^2) will
form an effective membrane within a few minutes of low pressure
pumping.

The subject of colloidal membranes leads into the role of
additives added to the feed in conjunction with the use of normal
film membranes. It has been known for several years that poly-
(vinyl methyl ether), various poly(oxyethylene-oxypropylene) copolymers based on ethylene diamine (trade name: Tetronic series),
as well as other additives improve the salt rejection of cellulose
acetate membranes without greatly lowering the flux of

water.[9,46,47] The rejection of urea from water can also be improved.[46] Evidently the minute amounts (milligram quantities as with colloidal membranes) of additive form a colloidal membrane superimposed on the film membrane, acting in effect as two membranes in series, or perhaps plugging defects in the base material. These comments should not imply that all slimes, colloids, particulate matter, or adventitous coatings are beneficial to membrane performance. In fact most coatings on membranes during use foul the membranes, resulting in a decrease in performance. Various organic solvents and water soluble organic acids as well as enzymes are used with and without "pressure shocking" to remove such fouling layers and thereby rejuvenate cellulose acetate membranes to lengthen their useful life[9,48] – a very important factor in the economics of reverse osmosis.

IV. Other Aqueous Separations

Since sea-water is not simply aqueous sodium chloride and since brackish water is often relatively low in chlorides, early in the study of pressure permeation attention was paid to separating a range of inorganic aqueous solutions.[9,49,50] Although the relative desalination of inorganic salts does vary, the premier membrane material is still cellulose diacetate.

Whereas cellulose acetate and certain other membranes will have a sodium chloride rejection of 95 percent or higher, the finding of a membrane for the separation of water-soluble organics is much more difficult. This is due of course to the sorption by the organic membranes of organic species. Early data for a cellulose acetate membrane having a sodium chloride rejection of 75 percent, showed a rejection for simple organics such as ethyl alcohol, acetic acid, urea, acetone, butanone, acetaldehyde, hydrazine, and butyric acid of 15-30 percent.[51] Branching on the permeate molecule showed a marked effect, witness tert-butyl alcohol 64 percent, sec-butyl alcohol 34 percent, iso-butyl alcohol 30 percent, and n-butyl alcohol nine percent. If the sorption of the organic is high enough, a negative rejection (enrichment) results as with phenol, + 20 percent[52] – again with cellulose acetate.

Although the removal of small organic molecules from water is difficult with cellulose acetate, larger molecules such as the tri-saccharide raffinose[53] or larger are removed in high enough degree, 75-90 percent, to make "reverse osmosis" a suitable step for the tertiary treatment of sewage or pulping liquors. Soluble "color bodies" are removed to the extent of about 80-95 percent with the standard cellulose acetate membranes. At this writing 11 cities in the U. S. are listed as studying or installing reverse osmosis units for municipal sewage treatment.[54] Some illustrative inorganic and organic rejections are shown in Table II. These are compounds associated with the photographic industry, reported by

Table II

Individual Rejections by EK RO-94 Cellulose Acetate at 600 psi and 27°C

Solute	Feed Conc. (g/l)	Percent Rejection	Water Flux (g/ft^2/day)
sodium sulfite	7	99	11
sodium thiosulfate	5	99	39
	200		7
sodium ferrocyanide	20	99	20
potassium ferricyanide	50	99	14
	200		8.5
sodium bromide	5	99	20
silver thiosulfate complex	0.1 Ag	99	38
potassium dichromate	20	94	-
hydroquinone monosulfate	10	85	-
formalin	27 ml/l	25	14
methyl p-aminophenol	6	18	10
hydroquinone	6	9	11
benzyl alcohol	2.5 ml/l	0	13

the maker of a commercial cellulose acetate asymmetric membrane in that same industry.[55]

Since the passage of aqueous solutions of organics through membranes by pressure permeation results in ameliorative rather than almost total separations, the only applications studied and published are in the waste treatment field where amelorative separations have some value. These studies which aim at optimizing the design of equipment and the gathering of economic data uniformly employ cellulose acetate membranes. While it would be of great value to dewater aqueous solutions of simple alcohols, glycols, aldehydes, and acids, no membranes now exist which differentiate enough in relative sorption to make such a separation possible by pressure activated transport. Simple sugars being highly hydrated are not as readily sorbed by organic membranes and are readily separated. Sugar beet refining and maple syrup concentration have been successfully demonstrated by at least three groups of workers on a pilot plant scale. Also easily separated are some surface active agents such as polyoxyethylated alkyl phenols.[56] Table III shows some early data by Sourirajan[57] for a preshrunk microporous cellulose acetate membrane on the removal of simple organics from water.

V.   Organic Mixtures

The interaction of an amorphous polymeric film and an organic penetrant can take on a variety of types of transport depending on temperature and activity of the penetrant, as illustrated in

Table III

Rejection by a Cellulose Acetate Membrane

Solute	Feed conc. (molal)	Pressure (psi)	Mole Percent Removal
sodium sulfate	0.5	1500	99
barium chloride	0.5	1500	95
lithium chloride	0.5	1500	91
sodium chloride	0.5	1500	83
ethyl alcohol	0.5	1500	36
n-propyl alcohol	0.5	1500	46
isopropyl alcohol	0.5	1500	67
n-butyl alcohol	0.25	1500	36
tert-butyl alcohol	0.25	1500	91
acetaldehyde	0.25	750	39
ethyl alcohol	0.25	750	27
acetone	0.25	750	24
acetic acid	0.25	750	21
glycerol	0.25	750	87
ethylene glycol	0.5	750	60
sucrose	0.5	750	99
dextrose	0.25	750	97
pentaerythritol	0.07	750	93
sodium formate	0.25	750	80

Figure IV, ideally in the lower left quadrant where the sorption is low and the energy of activation of transport is markedly less than 10 Kcal/mole. Here also diffusion is independent of concentration. In the upper portion of the diagram above the effective glass transition temperature (dashed line), the fluxes may be high but the semi-permeability will be low (e.g., elastomers). This means that the best membranes for separating a mixture of organics at room temperature will have an "ordered amorphous" structure with a rather high glass transition or softening temperature. That is they will be amorphous to x-ray diffraction rather than crystalline — for high flux, and ordered or "crystallizable" as viewed by electron diffraction for good semipermeability. Because organic liquids have a size comparable to the size of a monomeric unit in a polymeric membrane, more movement in the chain than the limited rotational oscillation necessary for gas transport is necessary for transport of the organic liquid. For this reason the glass transition temperature has more significance in the selection of membranes for organic than for gases. Although much theoretical work has been done especially in studying individual permeants,[59] relatively little has been accomplished in separating mixtures of organics by membrane processes. From a theoretical point of view these separations are a most difficult problem, since there is a high degree of interaction or coupling of effects. One simply cannot treat additively the

Figure IV

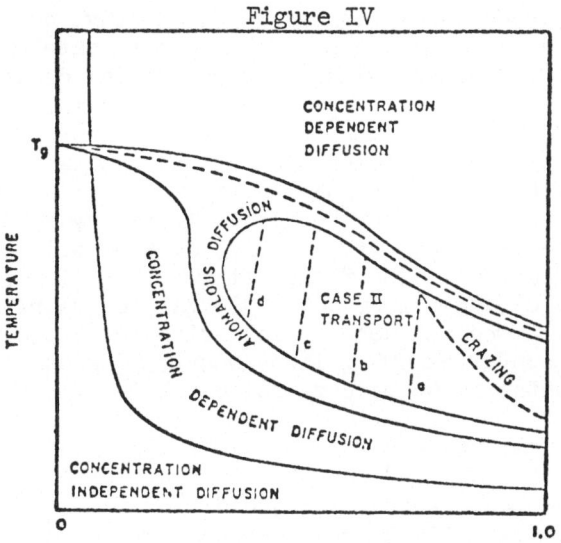

PENETRANT ACTIVITY

Transport features in the various regions
of the temperature-penetrant activity
plane for polystyrene-pentane.

effect of each component in the mixture on the membrane.

Since most of the organic liquids chosen for laboratory study
have low boiling points, the preponderance of work published on
separating organic mixtures by differential pressure has been performed by drawing a vacuum on the permeate side of the membrane
(downstream) rather than by imposing a pressure on the feed side
(upstream). The degree of separation achieved by this "pervaporation" (permeation + evaporation) is higher than that by pressure
permeation for two reasons. Firstly, the concentration gradient
of the mixture across the membrane is larger because one has a
divisor in the Nernst equation which is approaching zero on the
vacuum-pump side. Secondly, the morphology of the membrane itself
has more of a gradient. In pervaporation the upstream side is
swollen by the liquid feed while the downstream side is desiccated
by the vacuum pump or gaseous sweep. This constricted morphology
leads to more semipermeability than the case of pressure permeation
which has little or no gradient of sorption. Any increase in
sorption by the increased pressure on the liquid on the feed side
would be negated by the concomitant compaction of the membrane. Of
course one cannot use the pervaporation mode for mixtures containing
solid or high-boiling components.

In one of the few major studies reported ethyl cellulose, for-
maldehyde-modified ethyl cellulose, cellulose acetate butyrate, and
diisocyanate-modified ethyl cellulose were used as dense membrane
materials to enrich the following mixtures in the first component
by pervaporation:[60,61] heptane/isooctane, benzene/cyclohexane, and
methylcyclohexane/isooctane. The separation factors were in the
range of 2 to 9, this factor being the ratio of the components in
the permeate to their ratio in the feed. If the hydrocarbon mixture
does not swell the membrane sufficiently, an aromatic can be added
to increase the flux.[62] The fluxes at 700 Torr differential pressure at about 100°C ranged from 60 to 350 gallons/hour/1000 square
feet in this work. One run done at 165° with an unspecified cellulosic membrane converted 600 gallons/hour/1000 square feet of
47/53 toluene/naphtha to 66 percent toluene.[63]

In another published study of pervaporation binary combinations of hexane, ethanol, and acetone could be enriched in any
selected direction under unspecified conditions and fluxes depending on which of the following membranes were used: polyethylene,
polypropylene, rubber hydrochloride, copoly(vinylchloride-vinyl
acetate),[65] cellulose acetate, type 12 saran, and 300 MSD cellophane.

Using one component to prestructure a polyethylene membrane
favorably, some success has been achieved in separating xylene
isomers by pervaporation. Such a membrane preconditioned in the
para isomer shows an enrichment factor of 1.4 to 2.0 later with a
mixture.[66] Cold drawing a plasticized polyethylene is said to
increase the ortho isomer by a 1.8 factor.[67] These tailoring
effects can be made more permanent by irradiating the swollen membrane with gamma rays from cobalt 60. Both increases in flux and
semipermeability are claimed.[68]

Small separating effects have been generated by pervaporation
at low rate through membranes of microporous unfired Vycor glass.
The first component of the following combinations was slightly enriched: ethyl acetate/chloroform, ethanol/cyclohexane, methanol/
benzene, and ethanol/benzene.[69]

A comparison was made between unoriented and oriented polypropylene dense membrane for separating by pervaporation at temperatures between 25° and 75°C mixtures of the 50/50 benzene/
cyclohexane azeotrope and benzene/methyl alcohol.[70] Surprisingly
the oriented film showed a lower semipermeability with higher flux.
Separation factors were moderate, 1.25 to 2.4.

Although the above summary of published material on the vacuum-
driven pervaporation of organic mixtures may appear encouraging, in
fact, the authors know of no apparatus or trials for accomplishing
the separation of organic mixtures on other than a laboratory basis.

With the commercialization of "reverse osmosis" for the desalination of water, some attention has been paid to liquid organics.
The problem of predicting semipermeability to organic mixtures of
different types of membranes from their individual fluxes[71] has
been well described in phenomenological fashion.[72] Although
marked deviations from Darcy's Law are found in flux vs pressure
determinations for simple organics in crosslinked gum rubber, the
reciprocal of flux is linear with the reciprocal of pressure.[71]
This implies that there is a ceiling flux regardless of pressure
even in this membrane which is swollen four-fold. This would give
25 gal/ft^2/day toluene flux at 400 psi for a 1-mil rubber film.
Useful separations are carried out with an amount of swelling in
a much lower range, 15-25 percent, however.

Empirical data have been published from Sourirajan's laboratory on organic separations using cellulose acetate with a few
experiments employing polyethylene membranes. Studying annealed
commercial microporous cellulose acetate as well as asymmetric
cellulose acetate, low level separations were shown for ethanol/
xylene and xylene/heptane. Better results were achieved in separating the 2/1 ethanol/heptane azeotrope to a 90/10 composition, a
separation factor of 4.5. It was also shown that almost pure xylene could be realized at a much lower rate from 1/1 xylene-heptane
using a polyethylene membrane.[73]

More recently[74] extensive data has been published on the
separations of various alcohols from various hydrocarbons, separations among hydrocarbons, among alcoholic and xylene isomers, and
among alcohols. The permeability of the cellulose acetate membranes
was controlled by changing the annealing temperature in the range
from 78° to 98°C which decreased the standard ethanol flux 25-fold,
all at 1000 psi. Representative results are shown in Table V.

At least three points should be kept in mind by the reader
when considering the results shown in Table V. First of all this
represents experiments with an arbitrarily, preselected membrane
optimized for a totally unrelated separation — namely desalination.
Since cellulose diacetate has a Hildebrand solubility parameter of
about 11[75] while the permeants studied range in solubility parameter from 7.5 to 14.5; one could get better separations by changing membrane material to fit each mixture. Sourirajan, of course,
is aware of this as shown in his earlier paper.[73] Secondly, as
one changes the composition of an organic mixture, one is changing
the relative sorptions of the components for a given membrane.
For technological purposes one would want to optimize a membrane
for a given composition of a given mixture. Even within a run, to
the extent that the experiment is successful in forming a different
product from a given feed, the morphology of the membrane changes
finally reaching a steady-state gradient. Thirdly, the history of
a given membrane sample is most important in determining its

Table V

Some Organic Separations by Cellulose Acetate Membranes of Organic Mixtures at 1000 psig

Membrane Annealing temp. (°)	Ethanol flux g/hr/cm²	Feed	Product	α	Flux g/hr/cm²
90	4.5	MeOH/benz 62/38 azeo.	72/28	1.6	0.1
86	9	EtOH/heptane 2/1 azeo.	84/16	2.6	7.
86	7.5	EtOH/CCl$_4$ 40/60 azeo.	50/50	1.5	5.7
86	12.4	EtOH/cyclohexane 45/44 azeo.	3/1	3.6	4.2
87	6.2	BuOH/cyclohexane 1/10 azeo.	42/58	5.7	0.4
92	3.0	benzene/toluene 1/1	—	1.14	0.8
94	4.0	p-xylene/benzene 1/1	—	1.25	1.1
92	3.0	p-xylene/toluene 1/1	—	1.2	0.45
92	2.9	heptane/decane 1/1	—	1.2	0.12
98	0.75	n-PrOH/iso-PrOH 1/1	—	0.92(sic)	0.2
98	0.75	n-BuOH/isoBuOH 1/1	—	1.3	0.1
90	2.3	n-BuOH/tert-BuOH 1/1	—	1.13	0.2
90	2.3	tert-BuOH/sec BuOH 1/1	—	1.04(sic)	0.2
95	2.1	p-xylene/o-xylene 1/1	—	1.2	0.1
95	2.1	m-xylene/o-xylene 1/1	—	1.3	0.07

morphology, hence its semipermeability.[51,74] Every parameter in
the solubilization, casting, leaching, post-treatment, storage, and
use steps influences the morphology and therefore the characteristics of a given membrane material.

VI. The Separation of Gases

    Gaseous separation by pressure permeation offers three tre-
mendous advantageous not enjoyed by similar treatment of liquid
mixtures. Because the self diffusion of gas molecules on the feed
side of a membrane is usually quite high, there is normally no concentration polarization. That is, in a continuous flow system the
membrane is in contact with the bulk composition of the feed un-
perturbed by the flux through the membrane. Secondly, since the
interaction of inert gas molecules with the membrane material is
minimal and gas solubilities in membrane materials are low, membrane
morphology remains constant before, during, and after use.[59]
Thirdly, with most gases of practical interest there is no corrosion
problem.

    Although in the past 50 years since the introduction of re-
generated cellulose as a packaging film, an entire industry has
grown up based on the barrier properties of thin films, there has
been intermittent interest in membrane permeation. Even before
the advent of the asymmetric membrane there was serious attention
paid to gaseous separations. The separation of air,[76,77] the
separation of helium from natural gas, the separation of hydrogen
from coal gasification tail gas and refinery gases,[78,79,80] the
separation of ammonia from nitrogen and hydrogen, and the separation
of carbon dioxide from other gases[81,82,83] were all studied.

    The membranes used in this work varied, including rubber,
ethyl cellulose, polystyrene, polyacrylonitrile, and silicone
rubber. A rather thorough assessment of many membranes for the
helium from natural gas separation concluded that FEP Teflon was
the most desirable from the point of view of flux and helium/
methane ratio.[84] It must also not be forgotten that several
commercial plants have been built and operated using membranes of
a palladium-silver alloy for separating hydrogen from refinery
gases.[85]

    In order to give an idea of the relative permeability of a
variety of commercial polymers, the correlative scheme of the
Syracuse group of the mid 1950's is introduced. This group correlated permeability for various gases (i,k) in various polymers[86,87] with the general equation for permeability:

$$P_{i,k} = F(\text{polymer})\ G(\text{gas})\ \gamma(i,k) \qquad \text{where } \gamma\ i,k \approx 1.$$

F is a value for each polymer; G is a value for each gas, and gamma is an interaction factor which can be taken as unity. The F and G factors are given in Table VI. By definition the F values of Table VI are the nitrogen permeabilities for these polymers. Illustrating the utility of Table VI, the actual permeabilities of these polymers for carbon dioxide varies between 15- and 37-fold the nitrogen values, while the correlation (F·G) arbitrarily assigns the value 24.2-fold. Of course, the researcher looking for a useful application of membrane separations seeks the anomalies not the correlations. For example, molded polystyrene gives an oxygen/nitrogen value of 2.9, close to the correlation value of 3.8, but two commercial samples of biaxially oriented film give values of 7.8 and 8.9, a much more interesting figure for air separation. Some other semipermeabilities are shown in Table VII illustrating the importance of highly polar pendant groups.

Two technological events in the last ten years have reactivated interest in gaseous separation. One, the development of asymmetric casting procedures for membranes, has already been mentioned. The other was a means of producing membranes of silicone rubber and its block copolymers free from pinholes down to less than 0.25 microns thick. Since silicone rubber is such a permeable polymer (2500 on the scale of Table VI), these thin high-flux membranes, besides enabling hamsters to live under water, have a practical utility without being asymmetric. Suggested uses for this material are blood oxygenation, wound dressing, animal enclosures for biological experimentation, sensors for gas measuring instruments, and various gas separations.

The following items are commercially available at this writing: unbacked, single-backed, and double-backed dimethyl silicone membranes; 50/50 dimethyl silicone-carbonate block copolymers as film, resin, and three sizes of capillaries; 1/2 and 2/1 ratios of the block copolymer as resins; three sizes of membrane permeators for separating SO_2/air, NO_2/air, H_2S/air, NO_2/NO, oxygen/nitrogen, and ammonia or helium or hydrogen from other gases.[88]

Again because of its easily accessible asymmetry, cellulose acetate along with silicone rubber has received the most attention in the literature. Before the normally wet membrane can be used for gas separations it must be dried, for example, by freeze-drying.[89,90] In Table VIII are shown some data comparing dense cellulose acetate (so-called "lacquer" membrane) with the freeze-dried asymmetric versions of the same material.[91] Since the exact thickness of the semipermeable layer is in doubt, the reader should note the change from permeability to permeation rate in going from the dense to the asymmetric samples.

Hydrogen can be separated from carbon monoxide, carbon dioxide, nitrogen, or methane and other hydrocarbons by permeators based on hollow fibers of poly(ethylene glycol terephthalate) which have an

Table VI

Correlative F and G values at 30°C

Film	F value $(ccs/cm/cm^2/sec/cm\ Hg \times 10^{11})$
Saran	0.0094
Mylar	0.050
Pliofilm No	0.080
Nylon 6	0.10
Kel-F (Trithene)	1.3
Pliofilm FM	1.4
Hycar OR 15	2.35
Polyvinyl butyral	2.5
Cellulose acetate P-212	2.8
Butyl rubber	3.12
Methyl rubber	4.8
Volcaprene	4.9
Cellulose acetate (15% plasticized)	5.0
Hycar OR 25	6.04
Pliofilm P4	6.2
Perbunan	10.6
Neoprene	11.8
Polyethylene	19
Buna S	63.5
Polybutadiene	64.5
Natural rubber	80.8
Ethyl cellulose (plasticized)	84
Silicone rubber	2500

Gas	G-value
Nitrogen	1.0
Oxygen	3.8
Hydrogen sulfide	21.9
Carbon dioxide	24.2

inside diameter of 18 micrometers and an outside diameter of 36 microns. The separation factor for hydrogen from the heavier gases is 30; 500 psi pressure differential is recommended for industrial use of these 18-feet long, one-foot inside diameter units containing over 32 million fibers. The hydrogen product rate for one of these permeators is about one standard cubic foot per second.[92] The design of equipment and the mathematics of processing options have been worked out in great detail[93] for gaseous separations relating to industrial separations and air pollution.

Table VII

Separation Factors for Various Films[*]

Membrane	Gas Mixtures					
	He/CH_4	H_2/CH_4	He/N_2	H_2/N_2	He/O_2	He/H_2
polyacrylonitrile	60,000	10,000	3,750	625	-	6
poly(ethylene terephthalate)	264	162	-	-	35.5	1.6
poly(vinyl flouride)	166	68.5	-	-	45	2.4
cellulose, regenerated	400	-	-	-	48	-
poly(hexamethylene adipamide)	214	-	-	-	39	-
polystyrene	14.6	21.2	16	22	5.5	.73
ethyl cellulose	4.8	6.6	10.8	15	3.2	.73

* J. E. Jolley, U.S. Patent, 3,172,741 (1965).

Table VIII

Permeability of Dense Film and Asymmetric
Cellulose Acetate Membranes at 22°C

Gas	Dense Membrane (41µm thick)		Asymmetric Membrane Permeation Rate for 0.13µm(calc) "skin" cm^3/cm^2 sec-cmHg x 10^{-6}	Helium α
	Permeability $(\frac{cm^3-cm}{cm^2-sec-cmHg})$ x 10^{-12}	Helium α		
He	360	-	106	-
CH_4	14	21	3.4	31
O_2	-	-	7.1	15
N_2	14	21	3.1	34
Ne	240	1.5	19	5.5
Ar	32	11	3.7	28
C_3H_8	< 0.1	>3600	1.9	55

VII. Liquid Membranes

Up to this point we have considered amorphous solid membranes
with the proviso that some of the additives, slimes, or miscella-
neous material found in colloidal membranes may be highly viscous
liquids. The Eyring theory of liquids, derived from his theory of
absolute reaction rates, as well as a Stokes-Einstein hydrodynamic

approach to diffusion and the Sutherland-Einstein equation for
self-diffusion all lead to relation between viscosity and diffusion
constant of this form:

$$\eta \ D = n^{1/3} \ kT \ .$$

One might presume from this that permeation rates of gases through
thin membranes of monomolecular liquids might be higher than that
through highly viscous amorphous polymers. In fact, permeability
of gases through high-boiling organic liquids is lower or at best
the same as through most polymers, probably because of diminished
"free volume" in the liquid. Some representative data are shown
in Table IX.[94] Nevertheless, liquid membranes are a valid subject
for study and when some "chemistry" is involved, can show high semipermeability and good flux rates. One permeator unit containing an
immobilized liquid membrane for separating carbon dioxide from
other gases is now commercially available.[88]

Recently published work on liquid membranes for gas separations
has emanated from the same group which has been active in studying
silicone rubber membranes for the same purposes. Five years ago
they filed for a patent[95] setting forth several types of thin
liquid membrane. A layer of water backed by silicone rubber showed
a carbon dioxide/oxygen semipermeability of about 25 and a carbon
dioxide/hydrogen ratio of 20. Liquid diethylene glycol gave a carbon dioxide/oxygen ratio of 14. A one percent aqueous solution of
agar-agar in the form of a gel had the same semipermeability for
carbon dioxide/oxygen as did water (24.4), noticeably far higher
than the ratio of the G-value correlation for amorphous polymers
(6.8) shown in Table VI. A water-plasticized film of poly(vinyl
alcohol) gave a carbon dioxide/oxygen ratio of 36. Parenthetically,
the use of membranes highly plasticized with tributyl phosphate for
the separation of aqueous uranyl nitrate from other nitrates has
been shown.[96] The support for these liquid membranes may be
either silicone rubber or a microporous metal or polymer.

Sulfur dioxide may be separated from other gases by a liquid
membrane of the dimethyl ether of tetraethylene glycol supported
by microporous cellulose or microporous cellulose ether or
ester.[97] The microporous cellulose was made by caustic hydrolysis
of asymmetric cellulose acetate.

Higher semipermeabilities result when the carbon dioxide or
sulfur dioxide gas are ionized in the liquid layer. Then, since
permeability equals diffusion constant times solubility, the flux
rates go very high for the gas which is solubilized. The authors
term this "facilitated transport by carrier species" and report
carbon dioxide/oxygen ratios up to 4100 for the liquid membrane containing carbonate and bicarbonate ions.[98] Most of this work
evidentally focused on the life-support problem in isolated en-

Table IX

Gas Permeation Data for Some Liquid Membranes

Liquid	Helium Permeability (cc-cm/sec-cm^2-cmHg)	Helium/Methane Ratio	Oxygen Permeability (cc-cm/sec-cm^2-cmHg)	Oxygen/Nitrogen Ratio
glycerine	9×10^{-11}	10	4×10^{-11}	4
ethylene glycol	7×10^{-10}	3.5	2.7×10^{-9}	3.5
hexanetriol	9×10^{-11}	4	-	
carbitol acetate	1.7×10^{-11}	4	1.5×10^{-11}	3
6-methoxyquinoline	4×10^{-8}	1.1	1.5×10^{-10}	4
-nitroanisole	2×10^{-9}	1	1×10^{-9}	3
Aroclor 1254 (Dow)				
chlorinated diphenyl	7×10^{-9}	8	8×10^{-10}	1.6
dinonylnaphthalene				
sulfonic acid	2×10^{-9}	3	5×10^{-10}	1.4
isooctyl hydrogen				
phosphate	2×10^{-9}	1	2.5×10^{-9}	2.5
phenyl triethoxy				
silicate	2×10^{-9}	0.25	4.5×10^{-9}	2.8

vironments, but pollution abatement and helium extraction are also areas of potential application.

VIII. Perspective and Future Trends[99]

Most of the workers now engaged in the field of pressure permeation are taking the same few membrane materials along the route: membrane to element design to permeator design to plant design. They are perfecting techniques for meshing membrane, support, pressure vessel, pressure source, and hydraulic flow. Once designed, all the aspects must be studied for reproducibility, quality control, and lifetime with the goal of defining costs all along the line. A few in academe are looking for entirely new approaches to thin films or for entirely new membrane materials. Most of the effort is devoted to desalination, some to pollution control, some to gas separations.

What is needed at this time is a generalized theoretical approach to correlation between polymer morphology and polymer structure, so that permeability could be maximized within the various polymeric classes (hydrocarbons, esters, cellulosics, amides, etc.) and defined in physical terms. The keystone to this problem is an independent physical property which could be measured for each sample in lieu of the empirical approach of merely casting a film, putting it in a pressure cell, and running an actual permeation test.

Other future developments which one can anticipate are entirely new methods for making thin films such as: electrodeposition, _in situ_ polymerization and reliable _in situ_ casting so that a 200 Å coating could be put on a finished support element after the unit is mechanically in place. The problem of coupled interactions between components and membrane in the separation of organic mixtures will best be solved by discovery of a family of thermoplastic organic molecular sieves in the 5-20 Å range with highly limited swelling.

The neophyte researcher approaching a separation problem with membranes in mind first seeks high semipermeability. With increasing experience he realizes that materials of adequate semipermeability exist, the real problem is to get high enough flux (e.g., thin interfaces). When he starts to design and build actual working apparatus, he realizes that a limiting aspect of a membrane material is often mechanical strength. Undoubtedly we shall see progress in all three sectors in the 1970's.

IX.   Bibliography

1.   J. H. Mitchell, J. Roy. Inst. 2, 101, 307 (1831) as quoted by
      Stern, ref. 93.

2.   A. G. Horvath, U.S. Patent 1,825,631 (Sept. 29, 1931) filed
      Dec. 9, 1926.

3.   S. Loeb and S. Sourirajan, UCLA Engineering Report 60-60
      (1960); U. S. Patents 3,133,132; 3,133,137 (1964).
3a.   The term "asymmetric membrane" was coined by one of the present
      authors (HZF) in 1964 as a less ambiguous and more proper term
      than the then-used anisotropic membrane, which implies water
      flow in the plane of the membrane. In the ionophoretic mode of
      electrodialysis, water can flow in the plane of the membrane.

4.   S. A. Stern, U. S. Patent 3,332,216 (1967).

5.   F. L. Harris, Kaiser Engineers "Engineering and Economic
      Evaluation Study of Reverse Osmosis" Oakland, Calif. April
      1968.

6.   J. D. Bashaw and T. A. Orofino, "Hollow Fibers for Reverse
      Osmosis," Second OSW Symposium on Reverse Osmosis, Miami, Fla.,
      April, 1969. (Monsanto-Chemstrand, Durham, N. C.).

7.   "Saline Water Conversion Reports" of the Office of Saline
      Water, U. S. Department of Interior.

8.   R&D Reports of the Office of Saline Water.

9.   U. Merten, Editor, "Desalination by Reverse Osmosis," MIT Press,
      Cambridge, Mass., 1966.

10.   R. N. Rickles, "Membranes, Technology and Economics", Noyes
      Development Co., Park Ridge, N. J., 1967.

11.   C. E. Reid and E. J. Breton, J. Appl. Polymer Sci. 1, 133 (1959).

12.   C. J. Malin, M. E. Rowley, and N. G. Baumer, U.S. Patent 3,342,
      728 (1967).

13.   A. Turbak, editor "Membranes from Cellulose and Cellulose
      Derivatives" 157th National Meeting, American Chemical Society,
      Cellulose, Wood, and Fiber Chemistry Division, April 16-17,
      1969. Wiley-Interscience Applied Polymer Symposium No. 13,
      New York, 1970.

14.   Second Reverse Osmosis Symposium of the OSW, Miami, Fla. April
      20-25, 1969.

15. H. E. Podall, AIChE Meeting, San Juan, P. R., May 1970.

16. B. Baum, S. Margosiak and W. Holley at ref. 14. (DeBell and Richardson); OSW Conv. Report 1969-70, p. 92.

17. OSW Saline Water Conversion Report 1969-70, p. 115.

18. C. W. Saltonstall, PURAQUA Conference, Rome, Feb. 1968 and refs. 3 and 15 (Aerojet-General). OSW Conv. Report 1969-70, pp. 91 & 96.

19. C. W. Saltonstall and P. A. Cantor, OSW Conv. Report 1969-70, pp. 91 & 95; ibid, 1968 p. 110.

20. S. Manjikian, ref. 14, (Universal Water); OSW Conv. Report 1969-70, p. 96.

21. V. T. Stannett and H. B. Hopfenberg, OSW Conv. Report 1969-70, pp. 98-101.

22. M. E. Cohen, B. M. Riggleman, and P. D. Drechsel, ref. 13, p. 47.

23. L. T. Rozelle, J. E. Cadotte, and D. J. McClure, ref. 13, p. 61 and references cited therein p. 71; OSW Conv. Report 1969-70, p. 117.

24. J. E. Cadotte, L. T. Rozelle, R. J. Petersen, and P.S. Francis, ref. 13, 76.

25. R. L. Riley, H. K. Lonsdale, C. R. Lyons, and U. Merten, ref. 14, (Gulf-General); OSW Conv. Report 1969-70, p. 93.

26. R. Bloch, O. Kedem, and D. Vofsi, Polymer Letters $\underline{3}$, 965 (1965).

27. A. S. Hoffman, M. Modell, and P. Pan, J. Appl. Polymer Sci. $\underline{13}$, 2223 (1969); ibid, 14, 285 (1970); cf. ibid, 14, 1339 (1970).

28. A. Peterlin and H. Yasuda, OSW Conv. Report 1967, p. 131.

29. A. Sharples, G. Maconochie, and G. Thomason, ref. 14; OSW Conv. Report 1969-70, p. 102 (A. D. Little, Scotland).

30. OSW Conv. Report 1969-70, p. 104.

31. C. W. Saltonstall, et. al., OSW Conv. Report 1968, p. 123 (Aerojet-General).

32. OSW Conv. Report 1969-70, pp. 444-448 (General Electric).

33. OSW Conv. Report 1969-70, pp. 5-8 (Shell & Research Triangle Institute).

34. OSW Conv. Report 1969-70, p. 27 (Westinghouse).

35. The Desalination Report, Vol. 6, No. 19; May 7, 1970, Washington, D. C.

36. A. E. Marcinkowsky, K. A. Kraus, H. O. Phillips, J.S. Johnson, and A. J. Shor, J. Amer. Chem. Soc., 88, 5744 (1966); Desalination 1, 225 (1966).

37. K. A. Kraus, A. J. Shor, and J. S. Johnson, Desalination 2, 243 (1967).

38. A. J. Shor, K. A. Kraus, W. T. Smith, and J. S. Johnson, J. Phys. Chem. 72, 2200 (1968).

39. OSW Conv. Report 1969-70, pp. 107-111.

40. J. S. Johnson, K. A. Kraus, A. E. Marchinkowsky, H. O. Phillips, and A. J. Shor, U. S. Patent 3,449,245, June 1969.

41. J. R. Fryer, J. L. Hutchison, and R. Peterson, Nature 226, 149, (1970).

42. E. R. Brownscombe, H. F. Dunlap, L. R. Kern, and T. K. Perkins, U. S. Patent 3,283,813, November 1966 (filed September 1965), continuation-in-part of an application of June 1962).

43. E. R. Brownscombe and L. R. Kern, U. S. Patent 3,331,772, July 1967 (filed August 1965, continuation-in-part of an application of June 1962).

44. Private communication from Dr. W. K. Chen.

45. H. C. Savage, N. E. Bolton, H. O. Phillips, K. A. Kraus, and J. S. Johnson Water and Sewage Works, p. 102, March 1969.

46. OSW Conv. Report 1969-1970, pp. 111-113.

47. W. L. Short, R. T. Skrinde, and D. G. Newton, p. 188 in J. E. Flinn, editor "Membrane Science and Technology," Battelle Symposium, October 20-21, 1969. Plenum Press New York 1970.

48. OSW Conv. Report 1969-1970, pp. 471-474.

49. S. Sourirajan and T. S. Govindan, Ind. Eng. Chem. Proc. Design Devel, 5, 422, (1966).

50. R. Blunk, Report 64-28 UCLA Water Resources Center, June 1964
     C. A. 63, 2747d (1965).

51. S. Sourirajan Ind. & Eng. Prod. R&D 4, 201 (1965).

52. H. K. Lonsdale, U. Merten, and M. Tagami, J. Appl. Polymer
     Sci. 11, 1807 (1967).

53. Suggested as a standard compound for waste water treatment by
     one of the authors in 1964.

54. D. G. Stephan and R. B. Schaffer, J. Water Poll. Control Fed.
     42, 399 (1970).

55. J. G. Mahony, M. E. Rowley, and L. E. West, p. 203 in J. E.
     Flinn, editor "Membrane Science and Technology" Battelle
     Symposium October 20-21, 1969. Plenum Press, New York 1970.
     This book also treats complex organic mixtures generated by
     other industries such as food, pharmaceutical, paper, etc.

56. S. Sourirajan and A. F. Sirianni, Ind. & Eng. Prod. R&D 5, 30
     (1966).

57. S. Sourirajan, Ind. & Eng. Chem. Fund. 2, 51 (1963).

58. H. B. Hopfenberg, p. 16, in J. E. Flinn, editor "Membrane
     Science and Technology" Battelle Symposium, October 20-21,
     1969. Plenum Press, New York, 1970 and references cited
     therein; H. B. Hopfenberg and H. L. Frisch, Polymer Letters
     7, 405 (1969).

59. J. Crank and G. S. Park, editors, "Diffusion in Polymers"
     Academic Press, New York, 1968.

60. E. C. Martin, R. C. Binning, L. M. Adams, and R. J. Lee, U.S.
     Patent 3,140,256 (1964). R.C. Binning and J. M. Stuckey, U.S.
     Patent 2,958,657 (1960).

61. R. J. Lee and J. F. Jennings, U.S. Patent 2,960,462 (1960).

62. J. M. Stuckey, U. S. Patent 3,043,891 (1962).

63. R. C. Binning and R. J. Lee, Div. Pet. Chem. (ACS) 6, No. 2
     A-17 (1961).

64. R. F. Sweeney and A. Rose, Ind. Eng. Chem. Prod. R&D 4,248(1965).

65. V. N. Schrodt, R. F. Sweeney, and A. Rose, Div. Pet. Chem. (ACS)
     6, No. 2 A-29 (1961).

66. A. S. Michaels, R. F. Baddour, H. J. Bixler, and C. Y. Choo,
     Div. Pet. Chem. (ACS) 6, No. 2 A-35 (1961); Ind. Eng. Chem.
     Process R&D 1, 14 (1962); U.S. Patent 3,299,157 (1967).

67. H. J. Bixler and A. S. Michaels, Preprint 32nd AIChE,
     Pittsburgh, May 1964.

68. R. D. Siegel and R. W. Coughlin, Nature 226, 938 (1970); C&E
     News, p. 37, May 18, 1970.

69. K. Kammermeyer and D. H. Hagerbaumer, AIChE, 1, 215 (1955).

70. M. Kucharski and J. Stelmaszek, International Chem. Eng. 7,
     618 (1967).

71. D. R. Paul and D. M. Ebra-Lima preprint 23f AIChE Meeting,
     Washington, D.C., November 1969.

72. A. S. Michaels, pp. 157-195, "Membrane Processes for Industry"
     Southern Research Institute, May 1966, Birmingham, Ala.

73. S. Sourirajan, Nature 203, 1348 (1964).

74. J. Kopecek and S. Sourirajan, Ind. Eng. Chem. Process Des. Dev.
     9, 5 (1970).

75. J. L. Gardon, "Cohesive-Energy Density" Encyclop. of Polymer
     Science, Vol. 3, Wiley, New York, 1965.

76. P. Margis, German Patent 17,981 (August 7, 1881).

77. M. Herzog, U. S. Patent 307,041 (1884).

78. S. Weller and W. A. Steiner, J. Appl. Phys. 21, 279 (1950).

79. S. Weller and W. A. Steiner, Chem. Eng. Prog. 46, 585 (1950).

80. S. Weller, U. S. Patent, 2,540,152 (1951).

81. D. W. Brubaker and K. Kammermeyer, Ind. Eng. Chem., 46, 733 (1954).

82. K. Kammermeyer, U.S. Patent 2,966,235 (1960).

83. K. Kammermeyer, Ind. Eng. Chem. 49, 1685 (1957).

84. S. A. Stern, T. F. Sinclair, P. J. Gareis, N. P. Vahldieck, and
     P. H. Mohr, Ind. Eng. Chem., 57, 49 (1965).

85. R.B. McBride and D. L. McKinley, AIChE 55th Meeting, Feb. 1965.

86. V. Stannett and M. Szwarc, J. Polymer Sci., 16, 89 (1955).

87. C. E. Rogers, J. A. Meyer, V. Stannett, and M. Szwarc, TAPPI
     39, 741 (1956).

88. "Permselective Membranes," Bulletin 8685A, The General
     Electric Co., Schenectady, New York, February 1970; data sheets
     August 1970.

89. U. Merten, U.S. Patent 3,415,038 (1968); Canadian Patent 830,
     832 (1969).

90. J. P. Agrawal and S. Sourirajan, J. Appl. Polymer Sci., 14,
     1303 (1970).

91   P. K. Gantzel and U. Merten, Ind. Eng. Chem. Process Dev., 9,
     331 (1970).

92. 'Permasep' Permeator, duPont Co., Wilmington, Del. February
     1970.

93. S. A. Stern, "Gas Permeation Processes', in S. Loeb and R. Lacey,
     editors "Industrial Processing with Membranes", Plenum Press,
     New York, 1970.

94. H. Z. Friedlander and F-M Wang, unpublished data, 1968.

95. W. L. Robb and D. L. Reinhard, U.S. Patent 3,335,545 (1967).

96. R. Bloch in J. E. Flinn, editor "Membrane Science and Technology"
     Plenum Press, New York, 1970.

97. W. J. Ward, U. S. Patent 3,503,186 (1970).

98. W. J. Ward and W. L. Robb, Science 156, 1481 (1967); U.S.
     Patent 3,396,510 (1968).

99. H. Z. Friedlander, "Membranes" Encyclopedia of Polymer
     Science, Vol. 8, Wiley, New York, 1968.

TRANSPORT IN COMPOSITE REVERSE OSMOSIS MEMBRANES

H. K. Lonsdale, R. L. Riley, C. R. Lyons, D. P. Carosella, Jr.

Gulf General Atomic, San Diego, California

ABSTRACT

Reverse osmosis desalination membranes have been prepared by laminating a very thin film of a semipermeable material to a finely porous support membrane. Film thicknesses are typically 1000 Å. Structurally, these membranes are similar to the asymmetric cellulose acetate membranes first made by Loeb and Sourirajan. The composite membranes offer a choice of materials from which to prepare the desalination film and the porous support. Composites made by this procedure exhibit stable single-pass desalination of seawater with typical water fluxes of 5×10^{-4} g/cm^2-sec (10 gal/ft^2-day). The water flux through the composite with a given difference in water activity depends on the water permeability and the thickness of the thin film and the characteristics of the pores in the surface of the support. Water permeabilities have been measured in direct osmosis. When the distance between pores is large compared with the thickness of the thin film, significant reduction in flux can occur. The relationship between film thickness, the porosity and pore size of the support, and this flux reduction has been calculated, and the calculations compare favorably with observations made on composite membranes.

INTRODUCTION

A significant advance in the use of membranes for separation
processes came about with the introduction of the Loeb-Sourirajan
cellulose acetate membrane several years ago (Ref. 1). Because
of the unique combination of flux and selectivity characteristics
exhibited by this membrane, it has been introduced into a number
of separation schemes, the most important of which at this time
is water desalination by reverse osmosis (Ref. 2). Electron
microscopic examination (Refs. 3,4) and other measurements (Ref. 5)
demonstrated that this membrane is anisotropic, possessing a very
thin, essentially dense film or "skin" on one surface, the remainder of the membrane being a highly porous substructure that
serves as a support. The thin film is typically on the order of
0.1 to 0.2 μ thick, and the diameter of the pores in the substructure is of the same order of magnitude.

Because it is effectively very thin and essentially free of
imperfections, the Loeb-Sourirajan membrane permits both high
permeation rates and high selectivity. In reverse osmosis desalination, for example, water fluxes of 0.5 to 1.0 mg/cm^2-sec are
typical at a net pressure (applied hydrostatic pressure minus
osmotic pressure) of 40 atm, with reduction in the concentration
of a sodium chloride solution of 97% to 98%. The major limitations
to the wider application of this membrane are inadequate selectivity
for single-pass seawater desalination, membrane life, and a characteristic decline in flux with time at high pressure.

We have previously described (Ref. 6) a new approach to the
preparation of very thin supported films. In the present paper,
we examine the transport properties of these "composite" membranes
in detail, with emphasis on the properties of the thin film and
the effect of the pore structure of the support material on the
permeation through the composite membrane. The latter problem has
apparently not been examined in detail previously.

It has been shown that transport through cellulose acetate
membranes can be described by means of a solution-diffusion model
(Refs. 2,5), and this model will be used here. The water flux,
J_1, is given by

$$J_1 = \frac{D_1 c_1 \bar{v}_1 (\Delta p - \Delta \pi)}{RT \Delta x} \quad , \quad (1)$$

where D_1 = the diffusion coefficient of water in the membrane
(cm^2/sec),
c_1 = the concentration of water in the membrane (g/cm^3),
$\bar{v}_1$ = the partial molar volume of water in the membrane
(cc/mole),

Δp = the applied pressure (atm),
$\Delta \pi$ = the osmotic pressure difference across the membrane (atm),
R = the gas constant (cc-atm/mole-°K),
T = the absolute temperature (°K),
Δx = the effective membrane thickness (cm).

The salt flux, J_2, is similarly given by

$$J_2 = \frac{D_2 K \Delta \rho_2}{\Delta x} \quad , \tag{2}$$

where D_2 = the diffusion coefficient of salt in the membrane
(cm²/sec),
K = the distribution coefficient for salt between membrane
and solution [(g/cc membrane)/(g/cc solution)],
$\Delta \rho_2$ = the difference in salt concentration across the membrane (g/cm³).

The quantities $D_1 c_1$ and $D_2 K$ are permeability coefficients.

The salt rejection is defined by

$$S = (\rho_2' - \rho_2'')/\rho_2' \quad , \tag{3}$$

where ' and " refer to feed and product solutions, respectively.
In terms of the flux equations, salt rejection is given for dilute
solutions by

$$S = 1 - \frac{J_2}{J_1 \rho_2'} = \left[1 + \frac{D_2 KRT}{D_1 c_1 \bar{v}_1 (\Delta p - \Delta \pi)} \right]^{-1} . \tag{4}$$

EXPERIMENTAL

The composite membranes described herein were developed for
use in seawater desalination (Refs. 7,8). The thin films were
prepared from a high degree of substitution cellulose acetate
(Eastman Chemical Products, Inc., 432-130B, the 2.83-ester, referred to as CA$_{2.83}$) because of the very high semipermeability
($D_1 c_1/D_2 K$) of this material with respect to sea salts (Ref. 5).
The thin films, 400 Å to 8 microns thick, were applied to the
porous support by vertically withdrawing the support membrane
at constant speed from a dilute, filtered solution of CA$_{2.83}$ in
chloroform. These membranes will be referred to as "integral"
composite membranes. The apparatus and procedure have been

described previously (Ref. 6). The viscosity of the thin film solutions was determined with Cannon-Fenske viscometers. To prevent intrusion of the chloroform solution into the pores of the support membrane, this membrane was first spray-coated with a filtered solution of polyacrylic acid ("PAA," Rohm and Haas Acrysol A-1, molecular weight <50,000). Thin film thicknesses were determined by floating a known area of film off onto a water surface, drying, and weighing; a dry density of 1.3 g/cm^3 was assumed. Thin films could not be removed intact from composite membranes, and the thicknesses were determined indirectly by preparing films on glass and PAA-coated glass surfaces (from which they could be floated off) under the same conditions.

Some of the composite membranes were prepared by picking the floating thin film up onto the support membrane, which was brought up from beneath the water surface. Two such composites were prepared by layering three 400-Å-thick films onto a support membrane.

The porous support membrane was prepared by an adaptation of a technique that apparently was originated by Zsigmondy (Ref. 9). It was cast from a solution of cellulose nitrate and cellulose acetate dissolved in a mixture of good and poor solvents. The membrane was dried under conditions of controlled temperature and relative humidity. Details of the procedure are presented elsewhere (Ref. 10).

The dry membrane, the bulk of which contained fine interconnected pores, was typically 100 μ thick. Bulk porosity, as determined from density measurements, was 60% to 70%. The airdried surface of the membrane had a lower porosity. Electron micrographs of this surface on four typical materials are presented in Fig. 1. The pore characteristics were dependent on the composition of the casting solution and the temperature and relative humidity at which the membranes were cast.

The cellulose nitrate-cellulose acetate (CN-CA) porous support membranes were further characterized by means of a low-pressure "membrane constant," i.e., the flux of water per unit of applied pressure. These measurements were performed in a flowing test loop at 2.7 atm with freshly distilled water.

The composite membranes were tested in a high-pressure test loop (described in Ref. 5). A nominal 3.5 wt % NaCl solution was passed over the membrane surface by means of a positive displacement pump at an applied pressure of 102 atm. The brine velocity (150 cm/sec) was sufficiently high that boundary-layer effects were unimportant in the measurements (Ref. 11). Water flux and salt rejection were measured, the latter from the conductivity of feed and permeate solutions. Water flux results are expressed

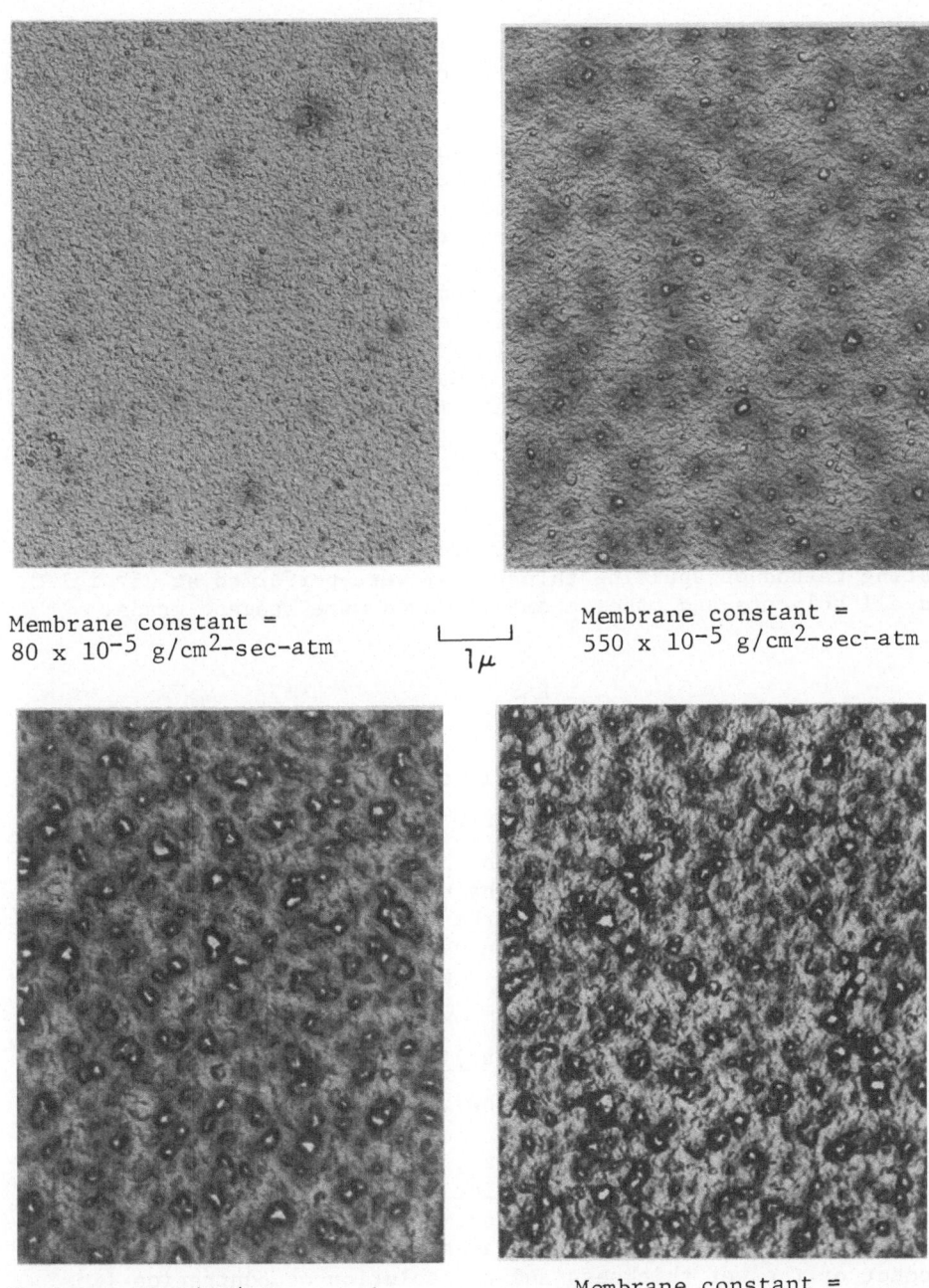

Membrane constant =
80 x 10^{-5} g/cm^2-sec-atm

$\vdash\!\!\!\dashv$
1μ

Membrane constant =
550 x 10^{-5} g/cm^2-sec-atm

Membrane constant =
4000 x 10^{-5} g/cm^2-sec-atm

Membrane constant =
6700 x 10^{-5} g/cm^2-sec-atm

Fig. 1. Electron micrographs of palladium-shadowed carbon replicas of the finely porous surface of several CN-CA support membranes. Material at the lower left is typical of that used to prepare composite membranes of Tables 2 and 3.

in terms of a membrane constant, the flux per unit net pressure
$[J_1/(\Delta p - \Delta \pi)$ in Eq. (1)], given in g/cm^2-sec-atm.

The water and salt permeabilities of "dense" films of $CA_{2.83}$
were determined in direct osmosis tests (Ref. 5). The films were
cast from a 5 wt % chloroform solution and allowed to dry in air
at 0% relative humidity. The area of film used was 260 cm^2. The
direct osmosis measurement was performed with the film separating
two halves of a Plexiglas cell. One half of the cell initially
contained 5.0 wt % NaCl solution, and the other half distilled
water. The conductivity of the latter solution was continuously
monitored by means of a Beckman Instruments conductivity cell with
a cell constant of 0.02 cm^{-1} and a Model RA4 Solumeter. The diffusion coefficient of salt was determined by the time-lag method
(Ref. 12) and the salt permeability (D_2K) from the steady-state
permeation rate and Eq. (2). The water permeability was determined
from the steady-state water flux and Eq. (1) (with $\Delta p = 0$).

All measurements discussed in this paper, including the dip-
coating method of applying thin films, were performed at 25° ± 1°C,
and all solvents and other chemicals used were reagent grade.

RESULTS

Thin Films

The thickness of the thin films was a function of the concentration of the solution from which the casting surface was withdrawn, the withdrawal rate, and the viscosity and surface tension
of the solution. The problem of the thickness of a liquid film
left on a smooth surface withdrawn vertically from solution has
been examined by Levich (Ref. 13). He found that the wet thickness is given by

$$\Delta x_w = \frac{0.93(\mu v)^{2/3}}{\sigma^{1/6} (\rho g)^{1/2}} \quad , \qquad (5)$$

where μ, σ, and ρ are the viscosity, surface tension, and density,
respectively, of the thin film solution; v is the withdrawal rate;
and g is the gravitational constant. The thickness of dry polymer
film left when the solvent evaporates is given, of course, by the
product of the wet thickness and the solution concentration (in
cm^3/cm^3). In most of this work, only dilute solutions were used,
and in substituting into Levich's equation, the surface tension
and density of the pure solvent (chloroform) were assumed. Thus,
σ = 26 dynes/cm and ρ = 1.50 g/cm^3. With these values, the predicted dry thickness, Δx, becomes

$$\Delta x = 0.016(\mu v)^{2/3}c \quad , \tag{6}$$

where c is the solution concentration expressed as g polymer/g
solution. The dependence of viscosity on concentration is shown
in Fig. 2. Observed thin film thicknesses are presented in Fig. 3
as a function of withdrawal rate. The lines drawn through the
data are the values obtained by means of Eq. (6). The agreement
is satisfactory, and the thickness is seen to increase with the
2/3 power of the withdrawal rate.

Porous Support Membrane

Porous CN-CA membranes were prepared under a variety of
experimental conditions. From electron micrographs of the finely
porous surface such as those shown in Fig. 1, approximate values
of the surface porosity, mean pore size, and distance between pores
were obtained. These results are summarized in Table 1. Typical
bar charts showing pore size distributions are presented in Fig. 4.
Porosities, ϵ, were estimated by integrating the pore area of these
bar charts, and the mean pore radius, r_1, was taken simply as the
value of the number maximum. The mean half-distance between pores,
r_2, was calculated by assuming that the pores were arranged in a
square array and taking the radius of the equivalent circle occupied
by each pore, i.e., $r_2 = r_1/\sqrt{\epsilon}$. The values of both r_1 and r_2 are
thus only crude estimates, but the lack of reproducibility in mem-
brane properties does not warrant a more elegant evaluation. The
increase in low-pressure membrane constant appears to be accompanied
by an increase in the size and perhaps also in the number of the
pores.

Direct Osmosis

Two measurements of the water and salt permeabilities were
performed. The steady-state water flow through a 20-μ-thick film
was 0.016 cc/hr, while that through a 19-μ-thick film was 0.023
cc/hr. The average water permeability calculated from Eq. (1) is
$8.2 \pm 1.9 \times 10^{-8}$ g/cm-sec. The steady-state salt flows were 4.3×10^{-5} and 8.3×10^{-5} g/hr, respectively, and the average salt permeability calculated from Eq. (2) is $2.7 \pm 1.2 \times 10^{-12}$ cm^2/sec.

Reverse Osmosis

Results of reverse osmosis experiments with six "integral"
composite membranes are summarized in Table 2. Membrane properties
tend to change slightly with time (i.e., water flux declines and
salt rejection increases), and the data in the table were collected

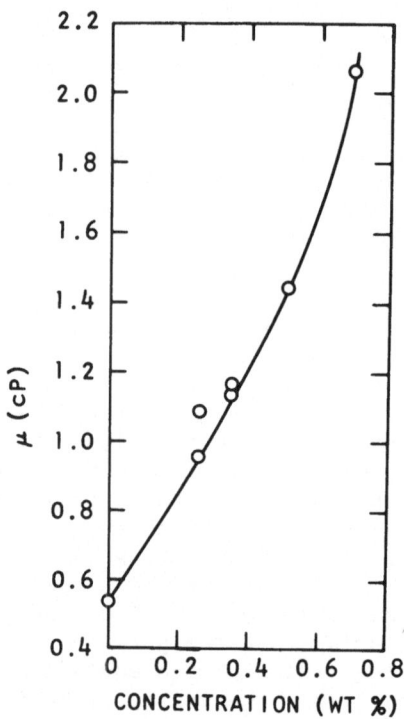

Fig. 2. Viscosity of 43.2%-acetyl cellulose acetate in chloroform
at 25°C

TABLE 1
SUMMARY OF PORE PROPERTIES OF POROUS SUPPORT MEMBRANES

Membrane Constant at 2.7 atm (10^{-5} g/cm^2-sec-atm)	Surface Porosity (%)	Mean Pore Radius r_1 (Å)	Mean Half-Distance Between Pores r_2 (Å)	Number of Pores per Square Micron
80	0.9	200	2100	7
550	6.6	400	2300	14
1400	8.7	450	1500	27
1600	9.2	500	1600	16
3500	14	600	1600	14
4000[a]	23	500	1000	24
6000	20	750	1700	21
6600	15	750	1900	22
6700	22	600	1300	26

[a] This is typical of the material used in the composite
membranes of Tables 2 and 3.

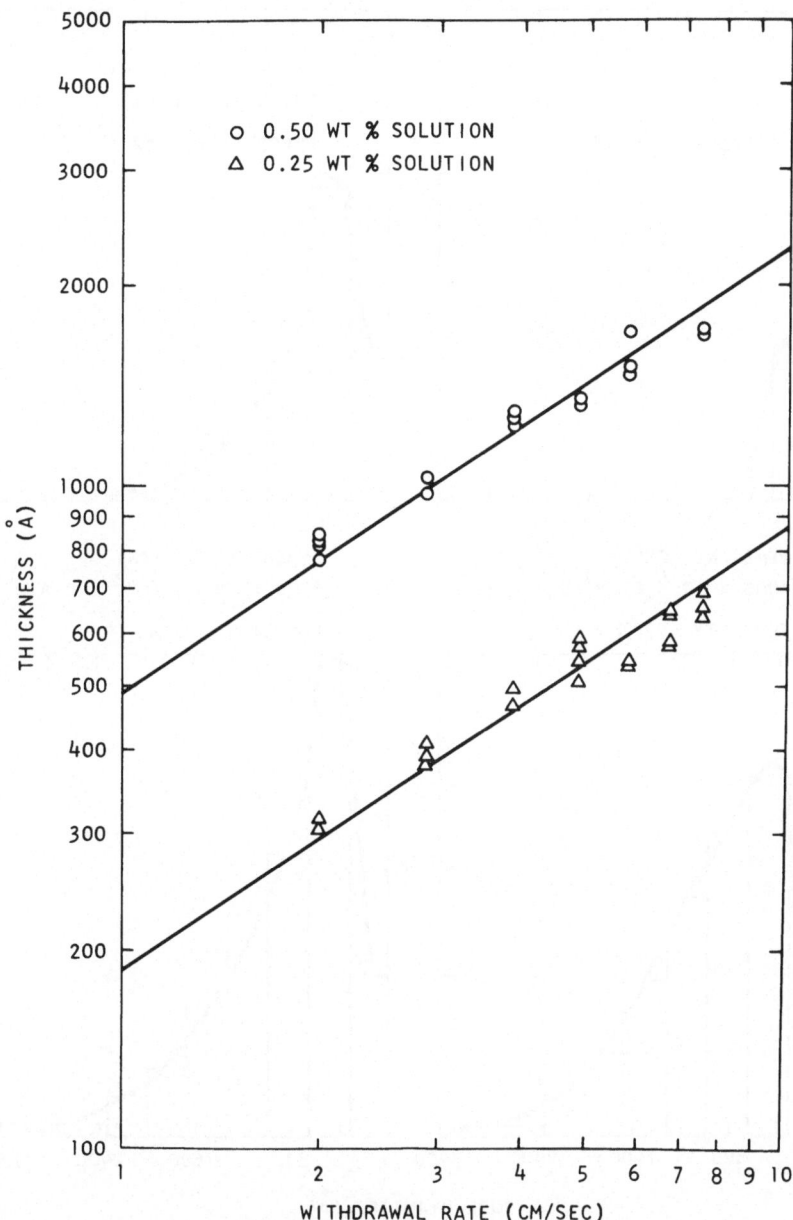

Fig. 3. Thickness of thin films of 43.2%-acetyl cellulose acetate
       vs withdrawal rate and concentration

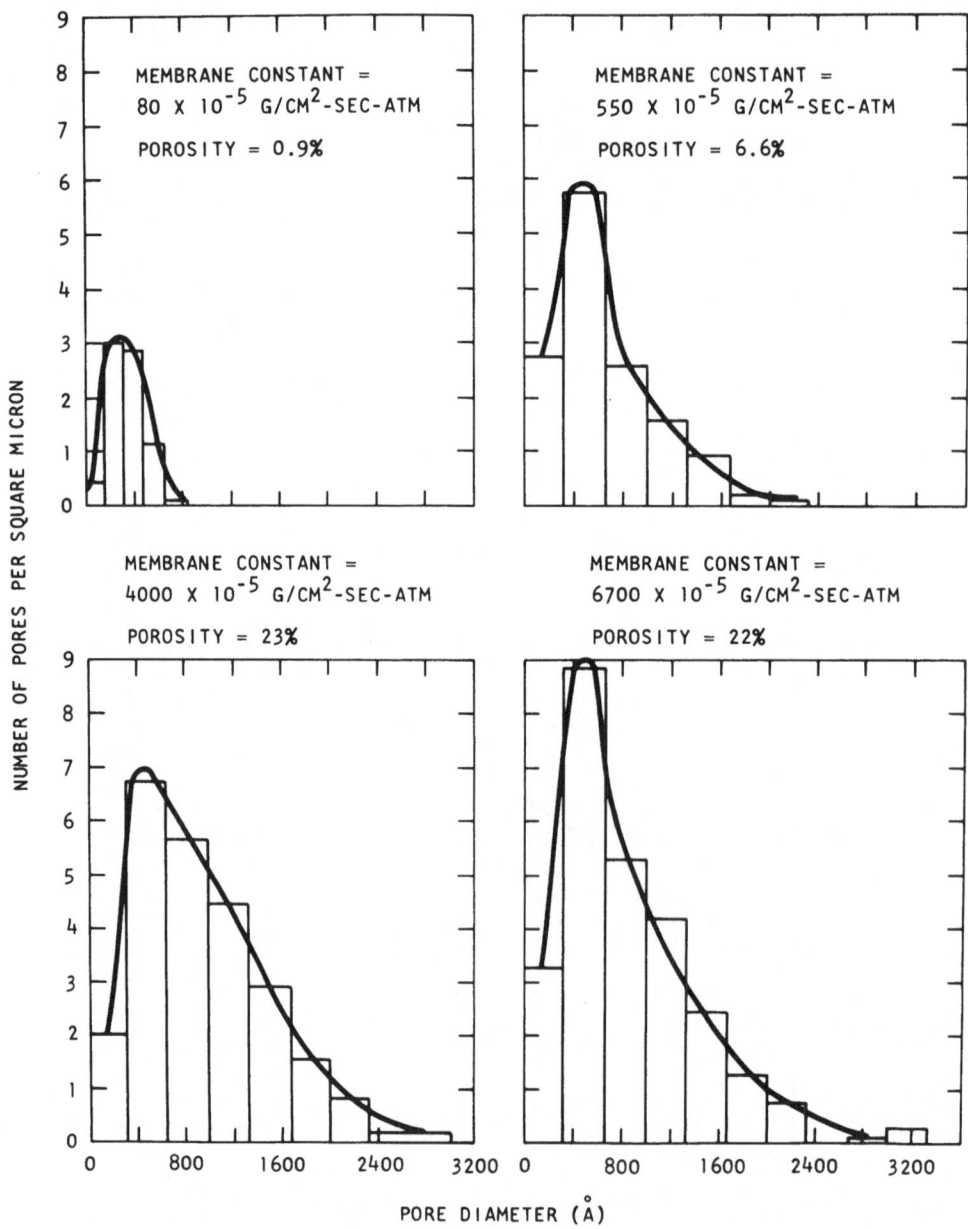

Fig. 4. Bar chart showing pore-size distribution of porous support
        membranes of Fig. 1

TABLE 2

REVERSE OSMOSIS RESULTS FOR "INTEGRAL" COMPOSITE MEMBRANES

Thin-Film Thickness (Å)	Water Flux (gal/ft2-day)	Salt Rejection (%)	Membrane Constant (10^{-5} g/cm2-sec-atm)	Observed Membrane Constant / Calculated Membrane Constant
68,000	0.12	99.87	0.0073	0.82
37,000	0.25	99.27	0.015	0.93
18,000	0.32	99.95	0.019	0.58
2,000	3.8	99.56	0.23	0.76
1,000	5.4	99.57	0.32	0.53
500	9.0	99.81	0.54	0.45

after the membranes had been on test for 48 hr. The feed solution
was 3.0 wt % NaCl. The ratio of observed-to-calculated membrane
constants was calculated from the water permeability value of
8.2 x 10^{-8} g/cm-sec and the known thickness of the thin film; this
ratio is discussed in detail below. The porous support used to
prepare these composites exhibited a membrane constant at 2.7 atm
of approximately 4000 x 10^{-5} g/cm^2-sec-atm.

Similar data for composite membranes prepared without the PAA
layer are presented in Table 3. The data were gathered after the
membranes were on test for 24 hr. The feed solution was 3.0 wt % NaCl.
The porous support used exhibited a membrane constant at 2.7 atm in
the range 4000 to 6000 x 10^{-5} g/cm^2-sec-atm.

Results of the reverse osmosis experiments performed with
composite membranes prepared with multiple thin films are presented
in Table 4. The feed solution was 3.5 wt % NaCl. The membrane
constant of the porous support at 2.7 atm was approximately 2500 x
10^{-5} g/cm^2-sec-atm.

DISCUSSION

Using the solution-diffusion transport model, it is possible
to calculate the expected water flux and salt rejection of the
composite membranes. For example, with 3.0 wt % NaCl ($\Delta\pi$ = 24 atm),
the salt rejection predicted from Eq. (4) for all the membranes in
Tables 2 through 4 is approximately 99.93%. Salt rejection should
be independent of thin film thickness, and indeed there is little
trend apparent over most of the range of film thickness. The
lower rejection apparent with some of the thinnest films is a
manifestation of the fact that the thin films failed over the
largest pores in the support membranes. Further evidence of this
failure is the fact that the sites of these imperfections were
decorated in the support membrane by means of a trace quantity
of ethyl violet dye present in the salt solution during the reverse
osmosis tests. However, most of the composite membranes with
thicker films also did not exhibit the expected salt rejection even
though, in some cases at least, no dye permeation was evident after
the tests. This is not believed to be a failure of the solution-
diffusion model but rather to be due to either undetected imperfections or history-dependent changes in the salt permeability of
the thin films. The fact that the salt permeability of cellulose
acetate films is quite sensitive to their method of preparation
and subsequent environment has been demonstrated; changes in salt
permeability of as much as an order of magnitude have been introduced, for example, with only slight change in water permeability
(Ref. 14). This would appear to be the rationalization for the
salt rejections exhibited by the composites with multiple thin
films (Table 4). One would expect that even if imperfections were
present in each of the three thin films, these would not overlap
and thus the adjacent film would effectively seal any imperfections.

TABLE 3

REVERSE OSMOSIS RESULTS FOR COMPOSITE MEMBRANES WITHOUT PAA

Thin Film Thickness (Å)	Water Flux (gal/ft²-day)	Salt Rejection (%)	Membrane Constant (10^{-5} g/cm²-sec-atm)	Observed Membrane Constant / Calculated Membrane Constant
78,000	0.12	99.33	0.0070	0.90
74,000	0.12	99.71	0.0072	0.88
36,000	0.22	99.50	0.014	0.84
34,000	0.24	94.4	0.015	0.84
18,000	0.35	99.87	0.021	0.64
16,000	0.42	99.90	0.026	0.70
2000	3.1	99.78	0.20	0.66
2000	2.8	99.53	0.17	0.56
1000	5.5	99.60	0.34	0.56
1000	5.5	99.47	0.34	0.56
500	10.0	98.7	0.60	0.50
500	8.9	98.7	0.55	0.45

TABLE 4

REVERSE OSMOSIS RESULTS FOR COMPOSITE MEMBRANES WITH MULTIPLE THIN FILMS

Number of Thin Films	Thin Film Thickness (Å)	Water Flux (gal/ft²-day)	Membrane Constant (10^{-5} g/cm²-sec-atm)	Salt Rejection (%)
3	400	5.4	0.33	99.49
3	400	5.2	0.32	99.76
1	1200	5.4	0.33	99.72

In spite of these discrepancies, it is important to note that the salt rejections are generally quite high and in most cases exceed those that can be attained with membranes of the Loeb-Sourirajan type. Those membranes can be "tailored" by annealing in water, for example, but even under the most favorable conditions, rejection of NaCl under the present conditions rarely reaches 99.5%. Most of the composite membranes exhibited salt rejections more than adequate to desalinate seawater to potable levels in a single pass.

It is also interesting to note that the PAA present during the preparation of some of the composite membranes does not appear to significantly affect their transport properties. In other experiments, analysis of the permeate for organic carbon content indicated that the PAA was not retained, but rather was swept through the porous support membrane in the first few minutes of the reverse osmosis test.

The expected water flux in the reverse osmosis tests was calculated from Eq. (1) using the thin film thickness and the water permeability measured on thick films in direct osmosis tests. The observed fluxes tended to agree with predictions for the thick films, but an increasing discrepancy developed with decreasing film thickness. This can be understood in terms of the model presented in Fig. 5. This sketch is an idealized magnification of the cross section of the thin film composite membrane. All the pores in the surface of the support membrane are apparently those visible in the electron micrographs; magnifications as high as 100,000x revealed no smaller pores. Further, comparison of observed low-pressure membrane constants with those calculated on the basis of the Poiseuille relationship and the size of the visible pores indicates that the nonporous region is present as a "skin" on one surface of the membrane which is several microns thick. The support membrane consists mainly of cellulose nitrate, a material with a much lower permeability to water than cellulose acetate. For these reasons, the nonporous regions of the porous support membrane can be considered impermeable, so that all the flux is constrained to pass through the pores. The net effect of this constraint is to increase the effective thickness of the thin film as indicated crudely in Fig. 5. The effect is small when the thickness is large compared with the distance between pores. However, as the thickness decreases the effect can become significant, and for $\Delta x \ll r_2$ the effective thickness approaches $\Delta x / \varepsilon$.

The reduction in flow due to the limited porosity of the support has been calculated by solving Fick's law with the appropriate boundary conditions. Referring to Fig. 6, these conditions are: (1) constant concentration everywhere at the upper surface of the thin film, (2) concentration equal to zero at the lower surface of the film everywhere over the pore, and (3) no flow across the interfaces indicated by the heavy lines in the figure. It was also

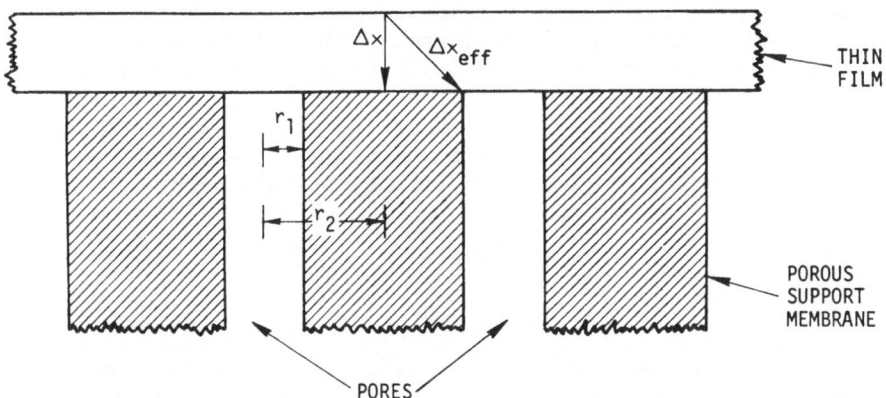

Fig. 5. Diagrammatic representation of flow path in composite membranes

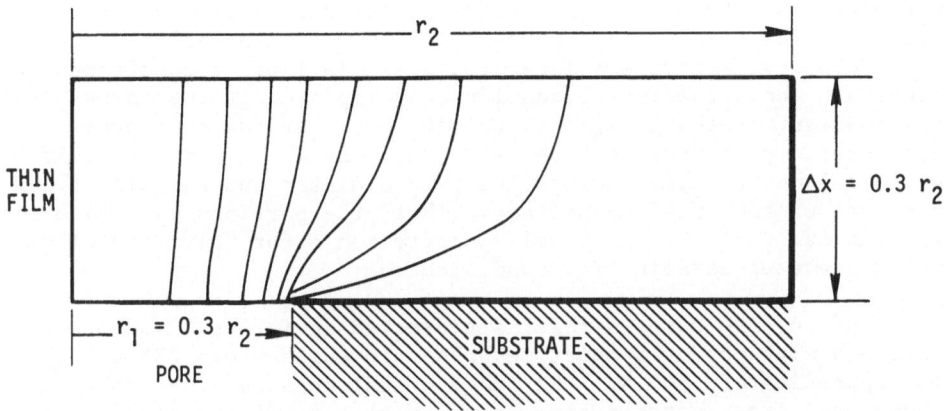

Fig. 6. Cross section of thin film over one-half of a pore, with flow contour lines. There is cylindrical symmetry.

assumed that the thin film was a homogeneous medium within which
the diffusion coefficient was constant. The calculation was
performed using finite difference methods on a Univac 1108 computer
by means of an existing two-dimensional heat-transfer code (Refs.
15,16). Because of geometric similarity, three parameters, Δx,
r_1, and r_2, define an entire family of cases. The results are
presented in terms of a geometric factor, G, defined by

$$G \equiv \frac{J_1(\text{actual})}{J_1(\text{ideal})} \quad ,$$

where J_1(ideal) is the water flux that would be observed with no
limitations due to porosity, i.e., the flux given in Eq. (1). In
addition, flow contour lines for a typical geometry were evaluated.
Because these lines are orthogonal to the lines of constant concentration, they could be obtained by reversing the boundary conditions. This set of lines is shown in Fig. 6 for the case of
$\Delta x = 0.3 \, r_2$ and $r_1 = 0.3 \, r_2$. Approximately equal mass is transported through each of the nine segments indicated.

Geometric shape factors are presented in Fig. 7 as a function
of the parameters r_1/r_2 and $\Delta x/r_2$. To illustrate more clearly the
effect of pore size, the results have been replotted in Fig. 8 in
terms of water flux [J_1(actual)] as a function of reciprocal thin
film thickness, $1/\Delta x$ (in microns^{-1}), porosity, ε, and pore radius,
r_1. Several factors are immediately apparent: the reduction in
water flux due to limited porosity can be substantial; for thin
films and relatively large pores, the flux asymptotically approaches
the "porosity limit," εJ_1(ideal); and with all other conditions
constant, the situation is considerably improved if the pores
are smaller. It should also be pointed out that for the large
hydrostatic pressures of interest here, many thin film materials
will fail mechanically unless the pore diameter and the film thickness are of comparable magnitude. Thus, the portions of Fig. 8
wherein $\Delta x \ll r_1$ are physically unattainable for polymeric films
under seawater desalination conditions.

The reverse osmosis results presented in Tables 2 and 3 were
obtained with porous support membranes in which ε was 23% and r_1
was approximately 500 Å. The range of thicknesses covered is
shown as a cross-hatched area in the plot for 20% porosity in Fig.
8. The column in Tables 2 and 3 labelled "Observed Membrane
Constant/Calculated Membrane Constant" now becomes interpretable
in terms of this model of restricted flow. A comparison between
the observed water flux and the calculated flux for (1) unrestricted
flow, i.e., a support porosity of 100%, (2) "totally restricted
flow," i.e., 23% of the unrestricted flow, and (3) partially restricted flow as calculated from Fig. 7 for the present conditions

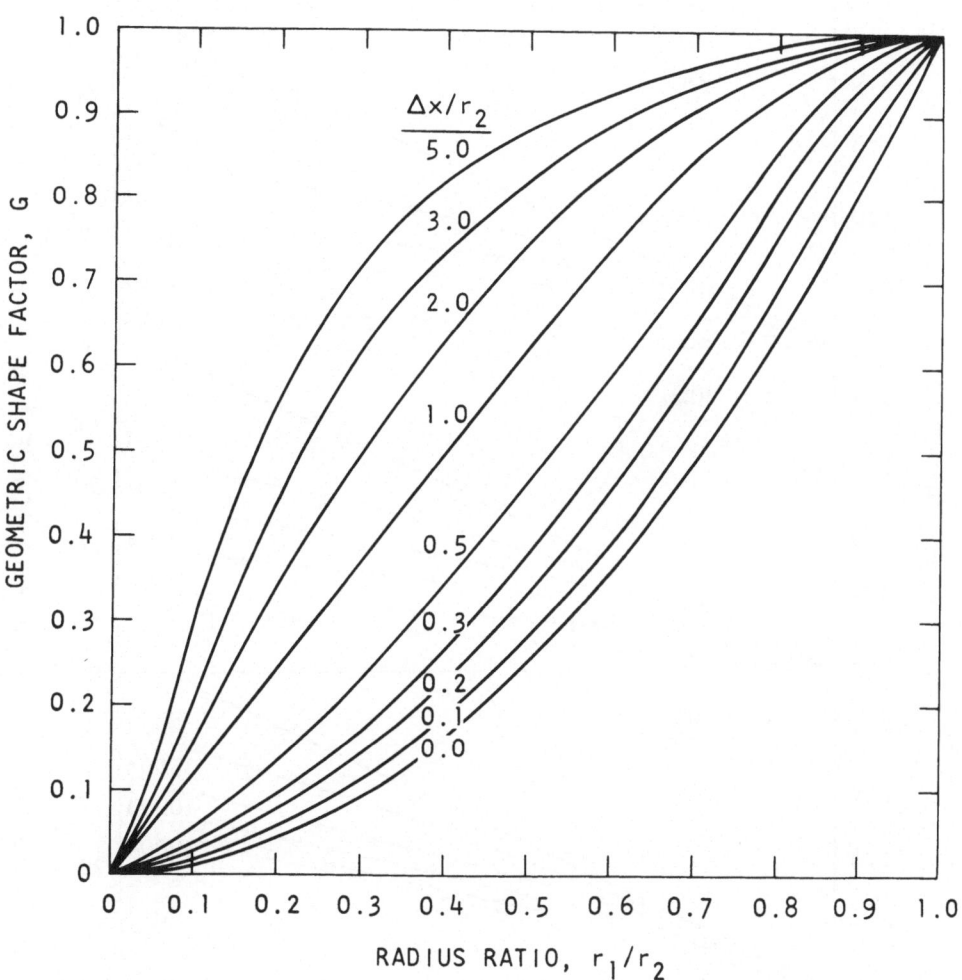

Fig. 7. Geometric shape factor vs radius ratio

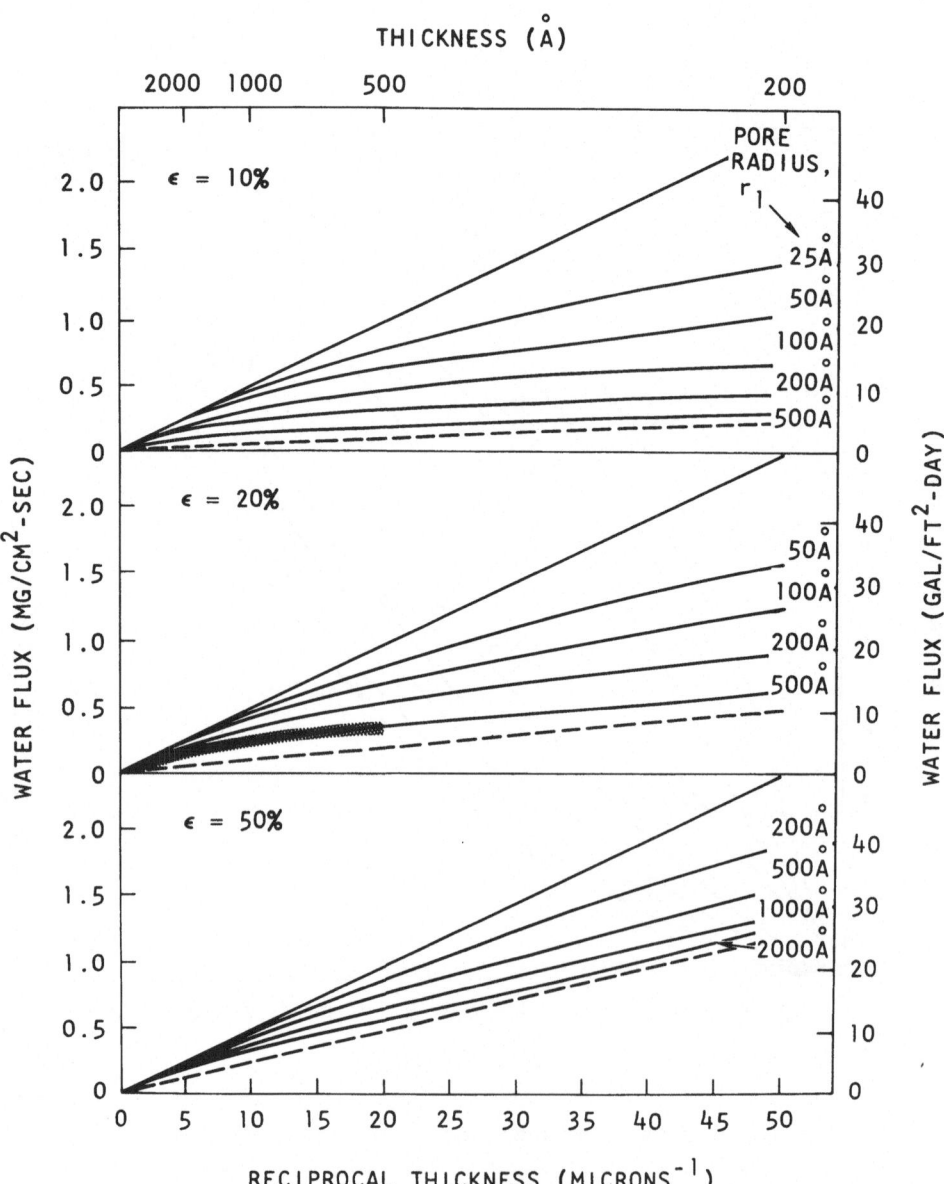

Fig. 8.   Water flux through composite reverse osmosis membrane vs
          thin film thickness and porosity and size of pores in
          support membrane

is presented in Fig. 9. The agreement between observation and the
calculation based on the restricted flow model is quite satisfactory
in view of the uncertainties in the assignment of porosity and pore
size.

In spite of the fact that the membrane constant of the support
membrane is orders of magnitude greater than that of the thin film,
it seems clear that limited porosity of the surface of the support
membrane reduces the water flux through the composite membrane.
The possibility also exists, however, that diffusion through the
thin film is not the rate-controlling step in the transport process,
but rather that sorption of water into the thin film at the high-
pressure interface and desorption at the low-pressure interface
are rate-controlling. Sorption-desorption kinetics are believed
to be rate-controlling in the transport of water across bimolecular
lipid films of 50 to 100 Å thickness (Ref. 17), and this mechanism has
been proposed to explain other transport data (Ref. 18). The fact
that the films used in the present study were considerably thicker
than lipid films might appear to preclude this mechanism here. The
results summarized in Table 4 confirm that this is the case. Each
of the composite membranes had a thin film 1200 Å thick, but in
those prepared with three 400-Å-thick films, it was necessary for
the water to be sorbed and desorbed three times. The fact that the
water fluxes through both types of membrane were essentially identical indicates that, at least at 400-Å thickness, the transport
process is diffusion-controlled.

CONCLUSIONS

The thin film approach described in this paper offers a new
route to the preparation of reverse osmosis membranes for desalination and, potentially, for other separation processes as well.
This membrane offers several possible advantages over the Loeb-
Sourirajan reverse osmosis membrane: in principle, thin films
can be prepared from a variety of materials; the thin film thickness can be reproducibly controlled in a straightforward manner;
and the porous support membrane can be prepared from materials
which are less subject to the compaction or flux-decline phenomenon
characteristic of the Loeb-Sourirajan membrane. The thin film
composite membranes exhibit excellent semipermeability and appear
to be virtually free of imperfections. Desalination of seawater
to potable levels is achievable in a single pass. In cellulose
acetate, the transport process is diffusion-controlled even with
films as thin as 400 Å. The pore properties of the surface of
the support membrane have been shown to be important to the water
flux. When the distance between pores is comparable to the thickness of the thin film, significant reduction in flux can occur.
The magnitude of the reduction has been evaluated quantitatively for
a wide range of film thicknesses and pore characteristics. The

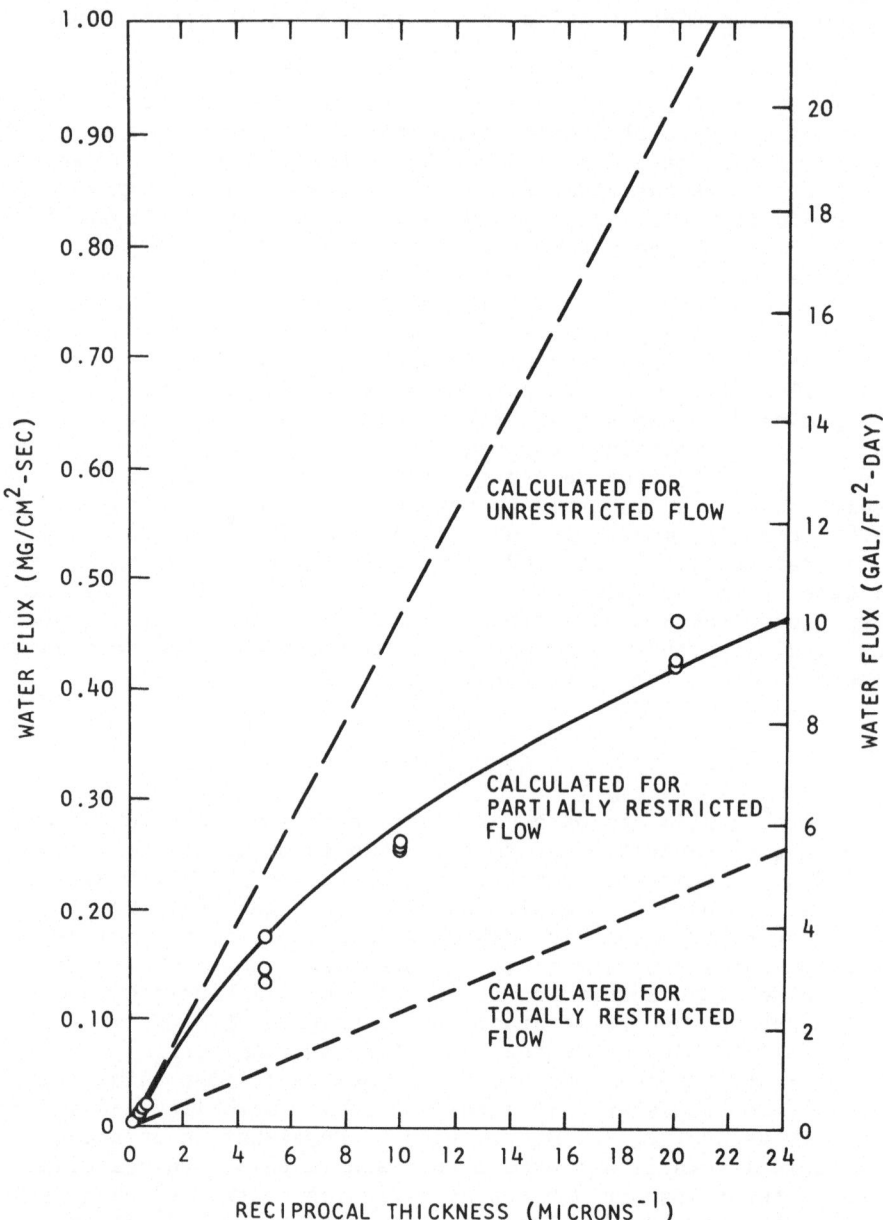

Fig. 9. Observed and calculated water flux vs reciprocal thin film thickness

use of films even thinner than those used in this study to enhance flux apparently cannot be tolerated. The films are already comparable in thickness to the pore diameter, about 1000 $\overset{\circ}{A}$, and thinner films fail at high pressure. Further improvements in membrane performance will require support membranes with higher surface porosity and finer pores.

The restricted flow model should have wider applicability. It will be useful in those cases of permeation through a supported, semipermeable barrier where the support structure has limited porosity, the nonporous regions are impermeable, and the spacing between pores is comparable to the thickness of the semipermeable barrier.

ACKNOWLEDGMENT

The authors are indebted to Mrs. G. Hightower and to D. Want for technical assistance and to Professor Vivian Stannett of North Carolina State University for suggesting the experiment described in Table 4. The electron microscopy was performed by S. Liang. This work was supported by the Office of Saline Water, U.S. Department of the Interior, under Contract 14-30-2609.

REFERENCES

1. Loeb, S., and S. Sourirajan, Advan. Chem. Series 38, 117 (1963).
2. Merten, U. (ed.), Desalination by Reverse Osmosis, The M.I.T. Press, Cambridge, 1966.
3. Riley, R. L., J. O. Gardner, and U. Merten, Science 143, 801 (1964).
4. Riley, R. L., U. Merten, and J. O. Gardner, Desalination 1, 30 (1966).
5. Lonsdale, H. K., U. Merten, and R. L. Riley, J. Appl. Polymer Sci. 9, 1341 (1965).
6. Riley, R. L., H. K. Lonsdale, C. R. Lyons, and U. Merten, J. Appl. Polymer Sci. 11, 2143 (1967).
7. Riley, R. L., H. K. Lonsdale, L. D. LaGrange, and C. R. Lyons, "Development of Ultrathin Membranes," U.S. Department of the Interior, Office of Saline Water, Research and Development Progress Report No. 386, Gulf General Atomic Incorporated, May 1968.
8. Lonsdale, H. K., R. L. Riley, L. D. LaGrange, C. R. Lyons, A. S. Douglas, and U. Merten, "Research on Improved Reverse Osmosis Membranes," U.S. Department of the Interior, Office of Saline Water, Research and Development Progress Report No. 484, Gulf General Atomic Incorporated, December 1969.

9. Zsigmondy, R., Z. Angew. Chem. 30, 398 (1926).
10. Lonsdale, H. K., R. L. Riley, C. E. Milstead, L. D. LaGrange, A. S. Douglas, and S. B. Sachs, "Research on Improved Reverse Osmosis Membranes," U.S. Department of the Interior, Office of Saline Water, Final Report under Contract 14-01-0001-1778, Gulf General Atomic Incorporated, April 1970.
11. Merten, U., H. K. Lonsdale, and R. L. Riley, Ind. Eng. Chem. Fundamentals 3, 210 (1964).
12. Daynes, H., Proc. Roy. Soc. (London), A97, 286 (1920).
13. Levich, V. G., Physicochemical Hydrodynamics, Prentice Hall, Englewood Cliffs, 1962.
14. Lonsdale, H. K., B. P. Cross, F. M. Graber, and C. E. Milstead, "Permeability of Cellulose Acetate Membranes to Selected Solutes," J. Macromol. Sci. Part B (to be published).
15. Clark, S. S., and J. F. Petersen, "TAC2D. A General Two-Dimensional Heat Transfer Computer Code - Mathematical Formulations and Programmer's Guide," USAEC Report GA-9262, Gulf General Atomic Incorporated, September 1969.
16. Petersen, J. F., "TAC2D. A General Purpose Two-Dimensional Heat Transfer Computer Code - User's Manual," USAEC Report GA-8868, Gulf General Atomic Incorporated, September 1969.
17. Tien, H. T., and H. P. Ting, Jr., J. Colloid Interface Sci. 27, 702 (1968).
18. Zwolinski, B. J., H. Eyring, and C. E. Reese, J. Phys. Chem. 53, 1426 (1949).

DEVELOPMENT OF DESIGN FACTORS FOR REVERSE OSMOSIS CONCENTRATION OF

PULPING PROCESS EFFLUENTS

I. K. BANSAL, GEORGE A. DUBEY, AND AVERILL J. WILEY

Chemical Engineer, Research Associate, and Chief;
Effluent Processes Group, The Institute of Paper
Chemistry, Appleton, Wis. 54911

ABSTRACT

Laboratory, pilot and large field demonstration units were used concurrently to develop engineering design factors for concentration treatment of four different pulp and paper industry waste streams by reverse osmosis. Eleven, half-inch diameter tubular modules with cellulose acetate membranes were operated in parallel under closely controlled conditions of flow rate (to 5 1/2 gallons/minute per module); velocity (to 9 feet/second); concentration (tap water controls to 10% waste liquor solids); pH (2.5-7.5); temperature (20°C to 40°C); viscosity (0.8-2.0 cp); and time of continuous operation (to ten days/run). Detailed data obtained include turbulence (N_{Re}); viscosity; osmotic pressure; pressure drop; permeation resistance; and effect of reverse osmosis in terms of rejections of dissolved solids, BOD, COD, and Color (OD) for the individual wastes processed. Design factors developed from these data include pumping energy per unit volume; hydrolysis rates on cellulose acetate; concentration polarization; fouling rates by microbiological growth and other surface accumulations; and membrane compaction rates. Auxiliary information is being developed from these studies to provide a firm base upon which methods of pretreatment and of operation can be developed to maintain sustained operation with minimum membrane fouling and also methods and routines for systematic cleanups where required.

INTRODUCTION

Prior reports in this series (2,3,6,7) review the development design and operation of reverse osmosis equipment for laboratory, pilot and three large field demonstrations for concentration processing of pulp and paper effluents. Design factors in the early phases of these studies were often based upon incomplete analytical characterizations of the wastes and their effect on performance of membrane equipment. The resulting estimations for design purposes proved out remarkably well in setting up the laboratory, pilot, and fairly large field demonstrations at flow rates ranging to 60,000 gallons per day and more. However, areas of need for more exact analytical and design data have become apparent as the trials proceeded and experience with field conditions developed. Particularly, there has been need for more exact data covering changes in processing variables as concentration proceeds in the range of 0.1% to 10% dissolved solids.

This, the fifth paper in this series (2,3,6,7) deals mainly with the following design data on calcium- and ammonium-base acid sulfite, N.S.S.C. and second-stage kraft bleach effluent liquors:

(A) Determination of Reynolds Numbers.
(B) Pressure drop and pumping energy requirements.
(C) Determination of osmotic pressures.
(D) Determination of rejection ratios.
(E) Product flux rate-temperature relationship for calcium-base acid sulfite liquor.
(F) Effect of velocity on flux rates of Na- and Ca-base liquors.
(G) Microbiological fouling and membrane compaction.

Determination of Reynolds Numbers

One of the first objectives in these studies was to determine required degrees of turbulence and mixing necessary to minimize concentration polarization and fouling of reverse osmosis membranes. These measurements were developed as Reynolds numbers, applicable at various temperatures and concentrations of the different types of pulp liquors. This is one of the important factors of concern in avoiding membrane "fouling" and for maintaining high flux rates. Concentration polarization produces several effects detrimental to the membrane separation process:

(1) The osmotic pressure that must be overcome is that corresponding to the solids concentration at the membrane surface. This is true, since concentration polarization causes the effective osmotic pressure of the bulk of the solution. For this reason, the required operating pressure for the reverse osmosis cell is increased by the

polarization effect, and the pumping power requirements
will also be increased.

(2) The concentration polarization may have a detrimental effect upon the dissolved solids content of the product
water because the dissolved solids content of this permeate will increase as the solids concentration at the
membrane surface increases.

(3) The deterioration of the membrane may be hastened by increased solids content of the product water, and concentration polarization can aggravate this effect.

(4) Finally, excessive concentration polarization may cause
precipitation of solids at the surface of the membrane.

Therefore, it is important to reduce concentration polarizing
by maintaining turbulent flow across the surface of the membrane.
The calculation of Reynolds number (N_{Re}) as a measure of the degree
of turbulence requires the determination of the density and viscosity of the liquor.

The viscosities of the liquor were determined using an Ostwald
Viscometer, whereas the specific gravities were determined using a
Pycnometer. The pH of each sample of the liquor was adjusted between 4.0-4.5 with sodium hydroxide or sulfuric acid, depending on
the initial pH. Then Reynolds numbers for different liquors were
determined using the experimental values of the densities and
viscosities at different temperatures and percentage solids of the
liquor. The data for N_{Re} and temperatures are fitted to a straight-
line relationship, using the method of least squares at different
percentage solids of the liquor.

The results of Reynolds numbers for N.S.S.C. white water,
ammonium-, and calcium-base acid sulfite liquors, and second-stage
kraft bleach effluent liquors, are plotted in Figures 1 to 4[1] at
different percentage solids and temperatures of the liquors. The
accuracy of the results was determined to be within 2%.

Turbulent flow is considered to occur at Reynolds numbers
above 4000. From Figure 1 to 4, it is apparent that a velocity of
about 1.0 foot/second should be sufficient to produce turbulence at
all temperatures indicated for a solids concentration of:

(a) 2.0% or less of N.S.S.C. white water.

(b) 3.0% or less of ammonia-base acid sulfite and second-stage
kraft bleach effluent liquors.

(c) 4.0% or less of calcium-base acid sulfite liquor flowing
in a tube of 0.5 inch inside diameter.

---

[1] Figures 1 and 2 have been reproduced from Reference (7) for com-
parative purposes.

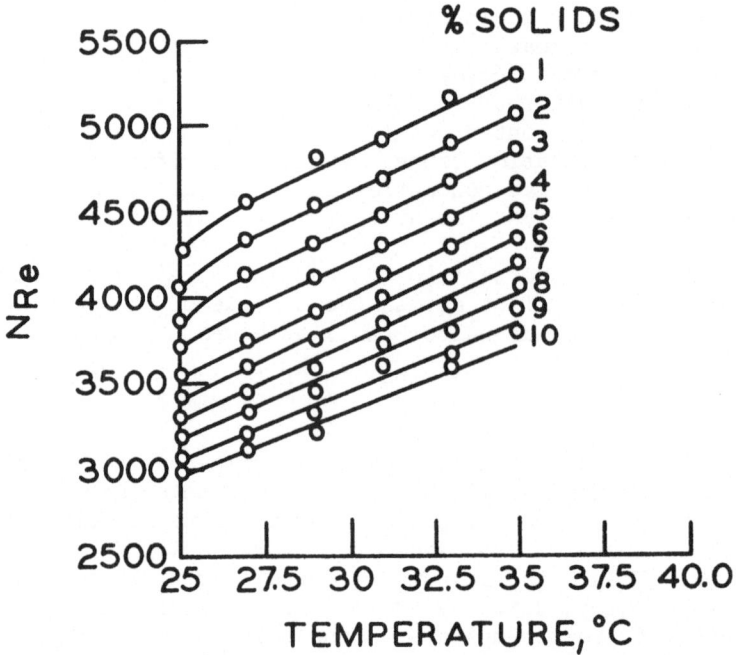

Figure 1:   Reynolds Numbers of N.S.S.C. White Water

For higher concentrations, a velocity of 1.0 ft./sec. may or may not be turbulent, depending on the temperatures of the liquor. The optimum and maximum desirable velocities are, of course, at much higher levels, which are discussed in Section F.

Pressure Drop and Pumping Energy Requirements for a Commercially Available Tubular-type Module[2]

A next step in developing engineering design data for reverse osmosis concentration processing of the four types of pulping and bleaching liquors involved determination of the pressure drop in a representative tubular module, and then to calculate the pumping

---

[2]Havens Model J 18 tube module having about 144 linear feet of 1/2-inch I.D. tubes in series.

energy as kilowatt-hours per 1000 gallons of liquor to overcome this pressure drop. This is one of the important design factors of concern in the selection of the number of modules to be used in series. If the velocity is to be held constant, then by connecting a large number of modules in series, we limit the total flow rate going into the system; but at the same time, the pressure drop increases in proportion to the number of modules. Under such conditions, it becomes necessary to add booster pumps to overcome the pressure drop. Therefore, one has to optimize pressure drop against the total flow rate, while selecting the number of modules to be connected in series.

Pressure drops in the 18 tube modules used in this study were determined at different flow rates for 4 liquors and water.

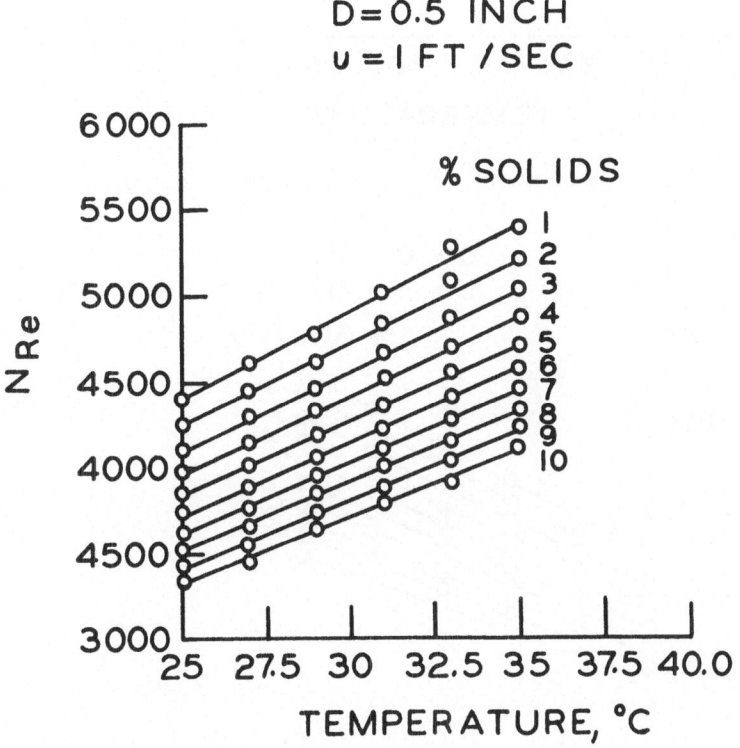

Figure 2: Reynolds Numbers of Ammonia-Base Acid Sulfite Liquor

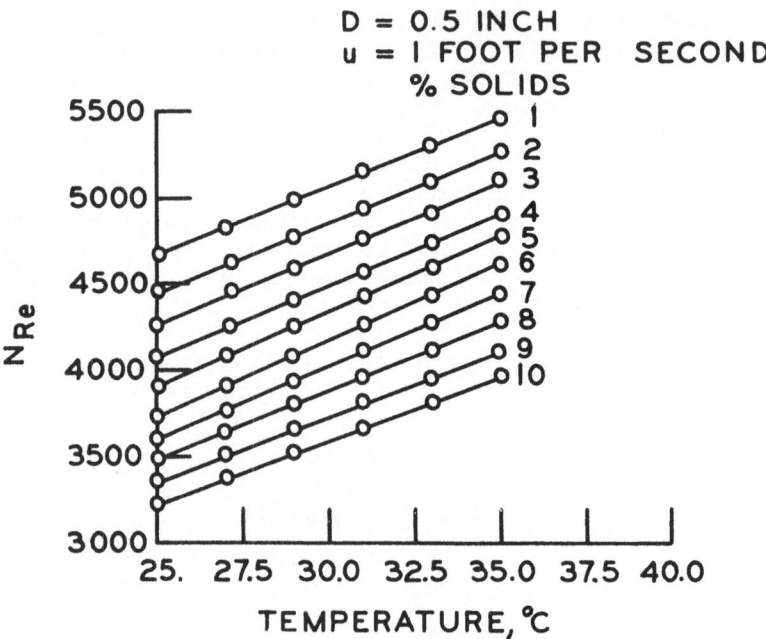

Figure 3: Reynolds Numbers of Calcium-Base Acid Sulfite Liquor

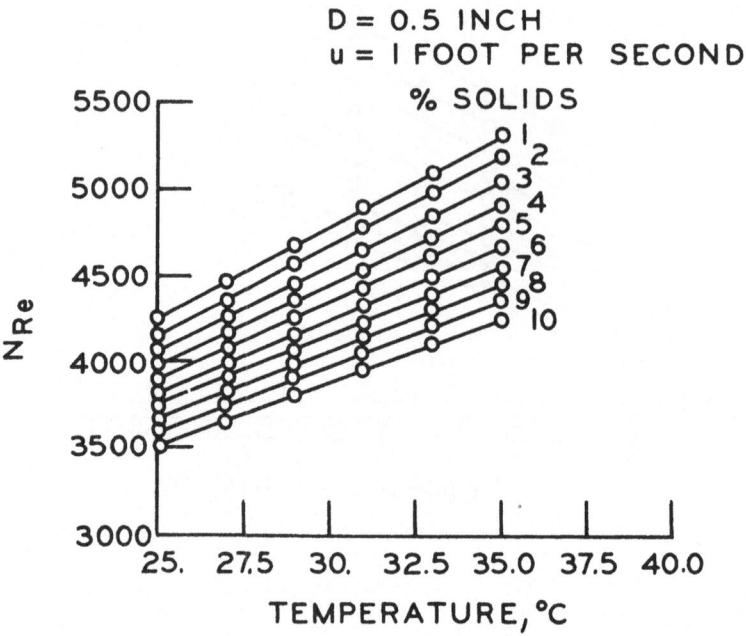

Figure 4. Reynolds Numbers of Kraft Bleach Effluent Liquor

Reynolds numbers, at different flow rates, were determined using the densities and viscosities of each liquor and water. Then the data for the pressure drops and N_{Re} were fitted to a log-log expression, using the method of least squares. Finally, the pressure drops at various values of N_{Re} were calculated from this expression, and the results are given in Figure 5 at 35.0°C for about 10.0% solids concentrations of 4 liquors and water. The pH of each liquor was adjusted to 4.5.

Pressure drop at N_{Re} of 40,000 is highest (= 166 psi) for N.S.S.C. white water, and lowest (= 94 psi) for kraft bleach effluent liquor. This is true because viscosities of various concentrations of kraft bleach effluent liquor were found to be lower than corresponding viscosities for N.S.S.C. white water at various concentrations.

From Figure 5, the pressure drop may be observed to increase rapidly with increase in N_{Re}. This is true because the pressure drop is directly proportional to $(N_{Re})^n$, where n varies from 1.75 to 2.00. The pressure drop in three identical 18 tube Havens modules connected in series was found equal to 3 times the pressure drop in a single 18 tube module. The higher the pressure drop, the lower the driving force, and the lower the flux rate. For example, in the case of three 18 tube modules connected in series, the pressure drop at N_{Re} = 40,000 for 103 g/l concentration of N.S.S.C. white water was 498 psi. So the average pressure of 351 psig for an inlet pressure of 600 psig can result in about 55% lower flux rates than the flux rates at 600 psig, even at 1.0% solids concentration of N.S.S.C. white water. Therefore, the pressure drop is a very important design factor in the selection of the number of modules to be connected in series.

Figure 6 gives the calculated pumping energy as kwh per 1000 gallons of liquor to overcome the frictional pressure drop at various values of N_{Re}. The pumping energy curves follow a pattern similar to the pressure drop curves. The higher the pressure drop, the higher the pumping energy. The highest value of pumping energy, which is for N.S.S.C. white water, at 103 g/l concentration, was calculated to be 1.72 kwh, compared to 0.56 kwh for water under the same values of N_{Re} = 40,000. Therefore, the kwh of pumping energy required for N.S.S.C. white water is 3 times that of water. The pumping energy is one of the important cost considerations in design and use of the reverse osmosis process.

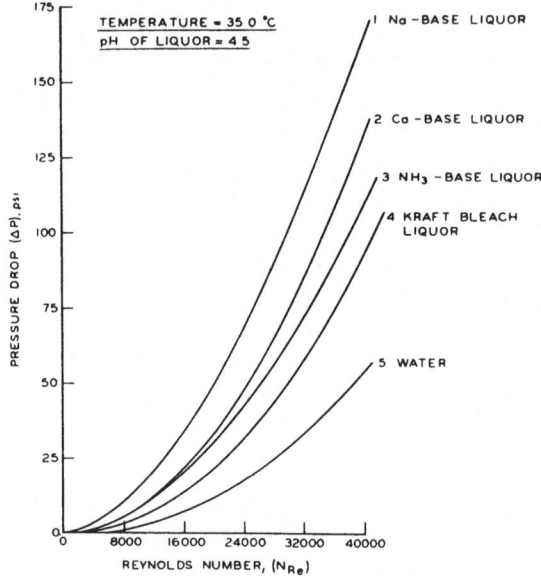

Figure 5:   Frictional Pressure Drop in One 18 Tube
Havens Module for 10% Solids Liquors

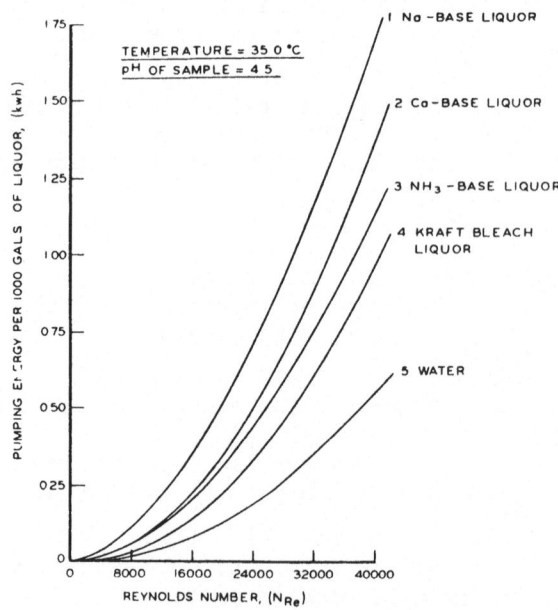

Figure 6:   Pumping Energy in One 18 Tube
Havens Module for 10.0% Solids Liquor

Determination of Osmotic Pressures

A next step in developing design data for effective reverse osmosis processing of these 4 liquors was directed to determination of the effective driving force required for different concentrations of each liquor. The flux rate at any concentration of the liquor is directly related to the osmotic pressure of the liquor by the following equation:

$$\underline{F} = \underline{A}(\Delta\underline{P} - \Delta\pi) \qquad\qquad (1)$$

where $\underline{F}$ = Flux rate through the membrane, gpd/sq ft

$\underline{A}$ = Membrane constant, $\dfrac{\text{gpd}}{\text{sq ft psia}}$

$\Delta\underline{P}$ = Difference between the applied pressure and the delivery pressure of product water, psia

$\Delta\pi$ = (Difference between the osmotic pressures of the liquor and the product water) + (osmotic pressure increase due to concentration polarization and fouling effects), psia.

The product water is delivered at atmospheric pressure. Since the osmotic pressure of the product water is usually very small compared to the osmotic pressure of the liquor, the former term can be ignored. In the case of zero concentration polarization and fouling effects, the driving force $(\Delta P - \Delta\pi)$ becomes equal to the difference between the applied pressure $(\underline{P}_A)$ and osmotic pressure of the liquor (π). Therefore, Equation 1 becomes:

$$\underline{F} = \underline{A}(\underline{P}_A - \pi) \qquad\qquad (2)$$

From Equation 2, it is apparent that the higher the osmotic pressure of the liquor, the lower the flux rate for a fixed applied pressure. In case of liquors having osmotic pressures higher than the applied pressure, there is osmotic flow across the membrane.

A vapor pressure Osmometer (V.P.O.), which operates on the principle of vapor pressure lowering, was used to measure the osmotic pressures of each sample of liquor. An NaCl solution was used as a reference for the V.P.O. However, the V.P.O. was useful only for determining the osmotic pressures of second-stage kraft bleach effluent liquor, and of the N.S.S.C. white water.

The osmotic pressures of calcium- and ammonia-base acid sulfite liquors could not be obtained by V.P.O., probably because of association and disassociation properties of lignosulfonates in these spent sulfite liquors, and were determined instead by measuring the flux rates at the different concentrations of each liquor. These liquor flux rates were then compared with the flux rates of sodium chloride solutions of known osmotic pressure. All the flux

rate runs were made at a higher velocity to minimize any increase
in osmotic pressure due to concentration polarization and fouling.
By solving Equation 2 and the sodium chloride flux rate equation,
$Fs = A(P_A - \pi)$, the following relationship is obtained for the
osmotic pressure of the liquor:

$$\pi = P_A\left(1 - \frac{\pi_s}{P_A}\right)\left(1 - \frac{F}{Fs}\right) \qquad (3)$$

where π = Osmotic pressure of the liquor, psia
P_A = Applied pressure, psia
F = Liquor flux rate, gpd/sq ft
Fs = Sodium chloride solution flux rate, gpd/sq ft
π_s = Difference between the osmotic pressures of sodium
chloride solution and the product water, psia.

By substituting the flux rates of sodium chloride and liquor
at 600 psig, and the osmotic pressures of sodium chloride in Equation 3, the osmotic pressures of calcium- and ammonium-base acid
sulfite liquors were determined. The results are given in Figure 7
at 25°C and at different concentrations of the 4 liquors.

From Figure 7, it is apparent that the osmotic pressures for
second-stage kraft bleach effluent liquor are very high compared to
other liquors, and it increases linearly from 84.0 psia at 10.0 g/l
to 594.0 psia at 100.0 g/l. This is understandable, because there
is more inorganic material, especially NaCl, present in second-
stage kraft bleach effluent liquor as compared to the other 3 liquors. The osmotic pressures of calcium-base acid sulfite liquor
and N.S.S.C. white water increase linearly with the concentration,
and are less than 300.0 psia for liquor solids concentrations up to
100 g/l. Ammonia-base acid sulfite liquor was found to have osmotic
pressures greater than for calcium-base acid sulfite liquor and
N.S.S.C. white water. In addition, the osmotic pressures of ammonia-base acid sulfite liquor do not vary linearly with the concentration.

Rejection Ratio

Dilute samples of each of the 4 liquors were concentrated
from 1.0 to 10.0% solids by reverse osmosis. During these concentration runs, feed, concentrate, and product samples were taken.
These samples were analyzed for solids, optical density (OD),
biological oxygen demand (BOD_5), and chemical oxygen demand (COD).
Rejection ratios (R) were calculated using the following formula:

$$R = 1 - Cp/Cc \qquad (4)$$

TABLE I.   REJECTION RATIOS (%) FOR 4 LIQUORS AT A
CONCENTRATION OF 0.2 — 10% SOLIDS

Inlet Pressure = 550-600 psig
pH of Liquor = 4.5 (Adjusted with H_2SO_4 or NaOH)
Inlet Velocity = 4-5 ft/sec
Type 3 Havens Modules

Sample		24-Hour neut. solids, g/l	Optical density at 281 nm	BOD_5, mg/l	COD, mg/l		24-Hour neut. solids, g/l	Optical density at 281 nm	BOD_5, mg/l	COD, mg/l
		Calcium-base acid sulfite liquor				N.S.S.C. white water				
1. Feed Sample		11.4	82	2820	13910		11.0	58	1870	6220
2. Concentrate Samples No.	1.	39.5	267	10740	48120	1.	57.1	298	9475	58200
	2.	91.6	675	24350	113680	2.	101.7	585	15900	127000
3. Permeate Samples No.	1.	1.23	3.9	864	1515	1.	1.47	1.3	854	1298
	2.	2.93	8.8	2078	3356	2.	1.95	1.8	822	1670
4. Rejection Ratios (%) No.	1.	96.90	98.54	91.96	96.85	1.	97.42	99.55	90.90	97.77
	2.	96.81	98.70	91.46	97.05	2.	98.08	99.70	94.84	98.69
		Ammonia-base acid sulfite liquor				Second stage kraft bleach effluent liquor				
1. Feed Sample		10.8	102	3210	13620		2.8	16	189	1375
2. Concentrate Samples No.	1	38.9	323	9650	49280	1.	19.4	117	955	11120
	2.	115.9	995	28050	143800	2.	86.1	642	4295	47840
3. Permeate Samples No.	1.	0.96	12.5	809	1420	1.	0.84	0.5	103	186
	2.	2.80	28.5	1910	3812	2.	5.13	1.1	256	413
4. Rejection Ratios (%) No.	1.	97.54	96.13	91.62	97.12	1.	95.65	99.60	89.21	98.33
	2.	97.59	97.14	93.19	97.35	2.	94.04	99.83	94.05	99.14

where C_p = Concentration of product
        $\underline{\bar{C}c}$ = Concentration of feed to the module.

The results are given in Table I at different concentrations of each liquor. Rejection of color based on OD measurements at 281 nm is good and ranged above 98.0% except in ammonia-base acid sulfite liquors. Ammonia-base sulfite liquors contain relatively low molecular weight colored materials, which are not completely rejected by Type 3 Havens cellulose acetate membrane. The rejection of solids is above 95.0% and range upward of that value in those liquors having good color rejections. BOD_5 rejections vary between 85.0 - 95.0%, and were observed to be significantly higher at upper levels of solids concentration for all types of spent liquors in these studies. COD rejections were found to be better than BOD_5 rejections and solids rejections, and were above 97.0% in most of the observations. The apparent anomaly with BOD_5 rejections being high at advanced stages of concentration, in some cases could be explained by permeation of low molecular weight organics, such as acetic acid, which are high in BOD_5 in early stages of concentration.

Product Flux Rate — Temperature Relationship

The effect of temperature on product flux rate was studied at 12.0 and 104 g/l solids concentration of calcium-base acid sulfite liquor. Figure 8 shows the percentage change is flux rate vs. temperature between 20.0 and 43.0°C. For calcium-base acid sulfite liquor of concentration 12.0 g/l, there was 2.1% increase in flux rate for every rise in degree centigrade of temperature. The percentage increase in flux rate did not change significantly at 104.0 g/l concentration of the liquor. Flux rate variation with temperature is higher for water and is about 2.8% rise per °C rise (6). According to Kopecek and Sourirajan (4), water flux rate increases due to increase in the membrane constant, which is related by the following expression at a given pressure.

$$\underline{A}/\mu\underline{w} = constant \tag{5}$$

where $\underline{A}$ = Pure water permeability constant, $\dfrac{g\text{-mole } H_2O}{cm^2 \text{ sec. atm.}}$

$\mu\underline{w}$ = Viscosity of water, centipoise.

According to Equation 5, the increase in flux rate with increase in temperature is a function of the decrease in viscosity of the solution. The larger the percentage change in viscosity of the liquor, the higher the percentage rise in flux rate. However, it is observed experimentally from Section A that the average decreases of viscosity per °C rise are 3.0% and 2.1% at 1.0% and

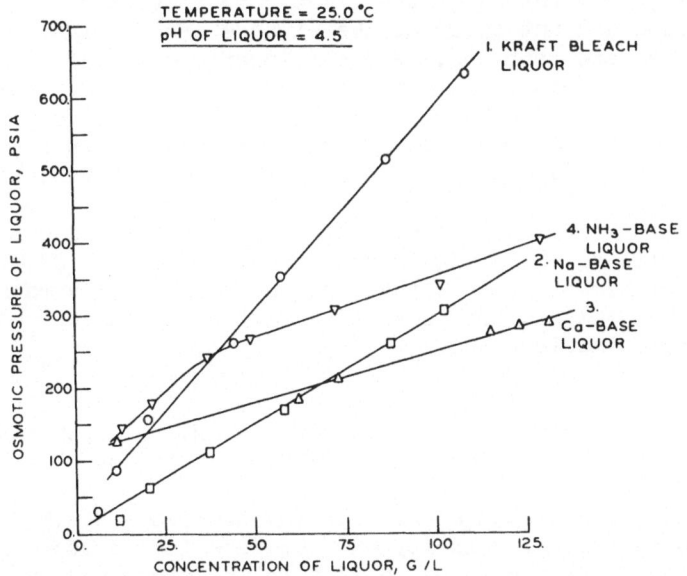

Figure 7: Osmotic Pressure vs. Concentration of the Liquor

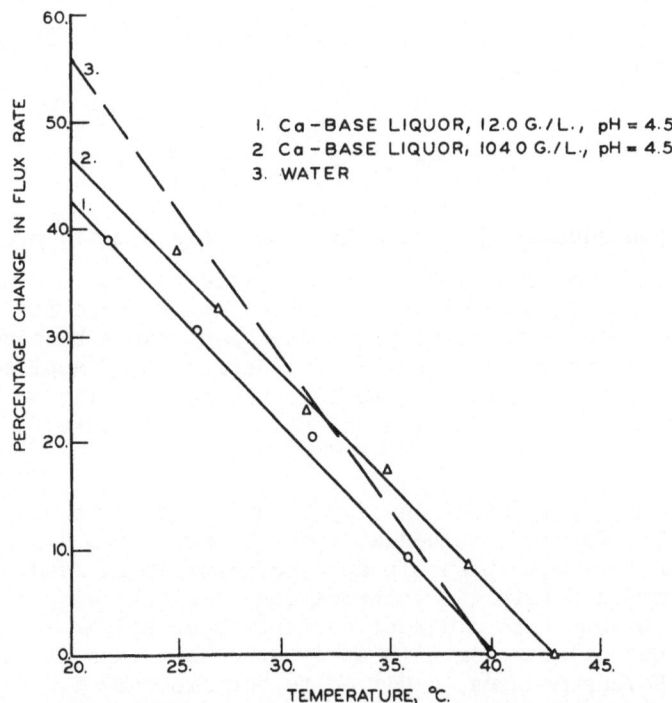

Figure 8: Effect of Temperature on Flux Rates of Water and Calcium Base Acid Sulfite Liquor

10% solids concentration of the liquor, respectively. This flux
rate increase did not occur at the expense of hydrolysis of the
cellulose acetate membrane. Hydrolysis of a membrane is a long-
term effect, whereas flux rate-temperature effect is instantaneous.
Of course, the rate of hydrolysis has been found to vary as the
reciprocal of the absolute temperature (5) and it can result in
increase in flux rate at the expense of percentage rejection.

Figure 8 shows that calcium-base acid sulfite liquor flux rate
at 40.0°C will be 25.0% higher than the flux rate at 28.0°C. Therefore, the higher the temperature, the higher the flux rate. We may
conclude that temperature is an important parameter in the design
of a reverse osmosis plant.

Effect of Velocity on the Flux Rates of N.S.S.C. White Water
and Calcium-Base Acid Sulfite Liquor

Permeation Resistance vs. Reynolds Numbers — N.S.S.C. White
Water. The effect of velocity on flux rate is expressed in terms
of permeation resistance, which in turn is determined by calculat-
ing the membrane constant. The membrane constant was determined
by using the osmotic pressure and flux rate data of a known solution
of sodium chloride. The permeation resistance (P_R) is calculated
using the following equation as derived and discussed in the section
on osmotic pressure determination:

$$P_R = P_A \left(1 - \frac{\frac{F_L}{F_s}}{}\right)\left(1 - \frac{\pi_s}{P_A}\right) \qquad (6)$$

The above equation, P_R, is the sum of the osmotic pressure of
the liquor and the osmotic pressure increase due to concentration
polarization and fouling effects. The effect of velocity on flux
rate is expressed in terms of permeation resistance because the
permeation resistance becomes almost independent of applied pressure and velocity when running sodium chloride and liquor flux
rates under identical conditions of velocity, pressure, and temperature.

A schematic diagram of the experimental setup is shown in
Figure 9. The feed was pumped at about 20 psig from a 500-gallon
plastic tank to a main piston pump by a centrifugal feed pump.
The main pump is a triplex reciprocating, positive displacement
Manton Gaulin pump with a direct current motor and an electronic
variable speed drive. For this study, the main pump discharged
through 11 Havens modules, modified to contain only two tubes at a
pressure of 500 psig and flow rate varying from 1.2-4.3 gpm in each

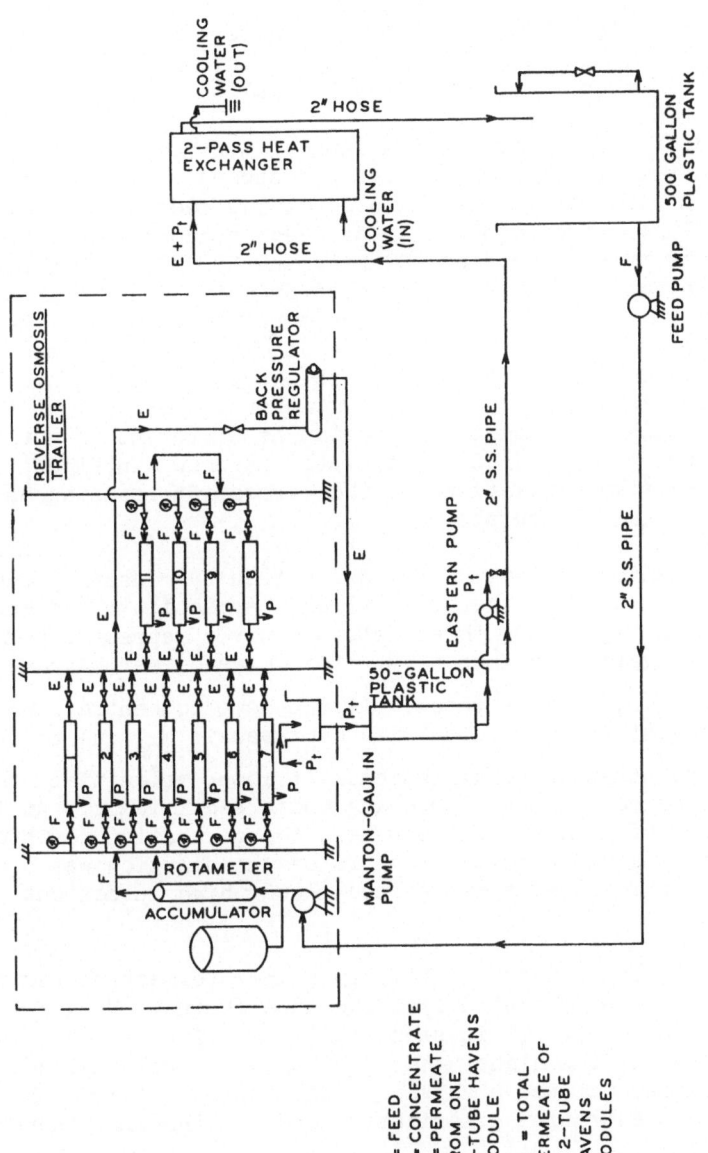

Figure 9: Schematic Diagram of Experimental Set-Up

module. A pressure gauge was installed at the inlet of each
module. The flow rate of permeate was measured from each module.
The total permeate of 11 modules was collected and mixed in a small
tank and then drained to a 50-gallon plastic tank underneath the
trailer. The permeate from a 50-gallon plastic tank was mixed with
the concentrate via a small centrifugal Eastern pump. The recom-
bined concentrate and permeate is returned through a heat exchanger
for cooling and then recycles to the 500-gallon feed tank.

Three different concentration runs of N.S.S.C. white water
were made at velocities 2.0-7.0 feet per second. The sodium chloride solution flux rates were measured at 5.0 feet per second for
each of the 11 two-tube modules before each concentration run.
Each flux rate run was made at 500 psig and 35.0°C. The results
of the permeation resistance calculated by using Equation 6 are
shown in Figure 10. Because of the uncertainty as to accuracy of
the osmotic pressures determined for sodium chloride as used in
Equation 6, there is a uniform error in the permeation resistances
and so the absolute values of the permeation resistances at the
highest velocities of each concentration is different from the
osmotic pressures of N.S.S.C. white water as shown in Figure 7. Of
course, the relative variations of the permeation resistances with
velocities are quite accurate.

The permeation resistance was determined by measuring the flux
rates of each module and then the average permeation resistance was
calculated. In Figure 10, three sets of curves represent permeation
resistances at various values of N_{Re} for three different concentrations of N.S.S.C. white water. At the same concentration, the
permeation resistance increased with decrease in N_{Re} due to increase in concentration polarization. At lower velocities, the
increase in permeation resistance was quite significant. At lower
concentrations of N.S.S.C. white water, the permeation resistance
was lower because the osmotic pressure of the solution was substantially less for that type of feed liquor high in content of
dissolved salts.

Velocities below which relatively higher permeation resistances
were obtained, were estimated from the data shown in Figure 10 and
are given in Table II. It is seen from Table II that the higher
the concentration, the higher the velocity below which higher
permeation resistance was observed. This is true because there is
increase in concentration polarization and fouling with increase
in concentration.

Concentration Polarization and Fouling Study as a Function of
Velocity for Calcium-Base Acid Sulfite Liquor. The effect of
velocity on flux rate is expressed in terms of percentage decreases
in flux rate from the starting flux rates. Here the same schematic

diagram of the experimental setup as shown in Figure 9 was used. A
number of continuous flux rate runs were made at different concentrations of the liquor under controlled conditions of 35°C and 500
psig pressure. Before the start of each run, the modules were
washed with high velocity water and with detergent BIZ solution of
15 g/l concentration. The percentage decrease in flux rate observed was less than 2.0% over a continuous run of 97 hours at 3.0 ft/
sec and for 118 g/l concentration of the liquor. As it became very
difficult to obtain fouling in case of two-module setup, we decided
to put nine 18-tube Havens modules in 3 manifolds, each manifold
containing 3 modules in series. The remaining 8 manifolds had 2
tube modules. The flux rate of each of the nine 18-tube Havens
modules were measured at different operating hours of the continuous run. The percentage decrease in flux rate from starting flux
rate was determined and then the average percentage decrease in
flux rate was calculated.

TABLE II. REYNOLDS NUMBERS AND VELOCITIES OF N.S.S.C.
WHITE WATER BELOW WHICH RELATIVELY HIGHER
PERMEATION RESISTANCES ARE OBTAINED

Concentration, g/l	N_{Re}	Velocity in 1/2" tube, ft/sec
12.7	16000	3.00
34.3	19000	4.00
56.0	22000	5.00

Figure 11 gives the average percentage decrease in flux rate
vs. hours of operation at two velocities of the liquor for concentrations above 10.0% solids of non-precipitated calcium-base acid
sulfite liquor. From Figure 11, it is noted that the average percentage decrease in flux rate at 70 hours of operation is reduced
from 8.0 to 4.0% by increasing the velocity from 0.8 to 1.2 ft/sec.

Figure 12 shows the effect of velocity on flux rate for calcium-
base acid sulfite liquor in which there was a significant amount of
precipitated calcium sulfate solids. The average percentage decrease in flux in this precipitated liquor is a strong function of
velocity and it becomes less than 12.0% at 1.8 ft/sec over 50 hours
of continuous operation. The conclusion is that velocity is an
important parameter in controlling the decrease in flux rate which
may result from scaling and fouling of membrane tubes.

Finally, it is noted that we have been able to control concentration polarization at velocities even below 1.5 ft/sec. Fouling
has not been apparent in 70-hour runs at these low velocities.
However, it is not economical to have such low velocities because
the absolute value of flux rate increases with increase in velocity

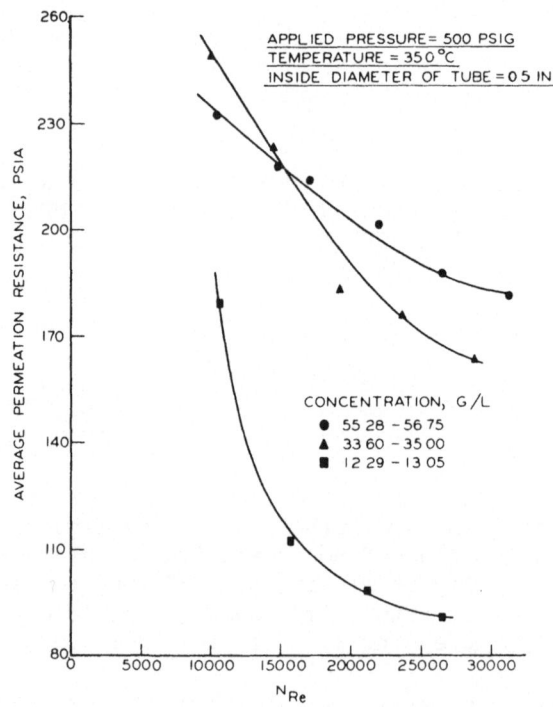

Figure 10: Permeation Resistance vs. Reynolds
Numbers — N.S.S.C. White Water

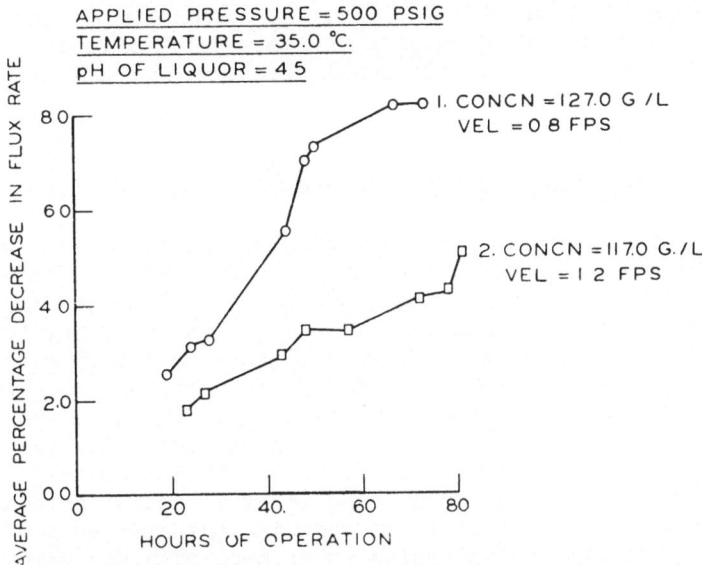

Figure 11: Effect of Velocity on the Flux Rates of Ca-Base Liquors (No Precipitate)

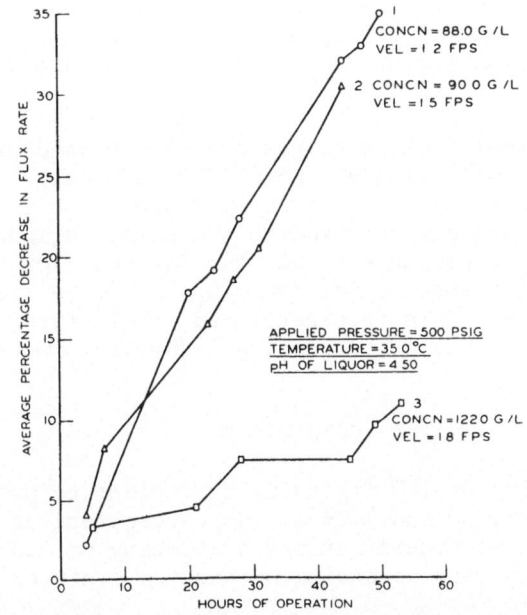

Figure 12: Effect of Velocity on the Flux Rates of Ca-Base Liquor (Precipitate)

at a rate proportional to $v^{0.8}$ (1). At higher velocities, it becomes necessary to optimize this increase in flux rate against the loss of flux rate due to frictional pressure drop and the cost of pumping energy.

Microbiological Fouling and Membrane Compaction

Fouling of the membrane surfaces by microbiological growth is often observed in sustained operations with wastes containing nutrients capable of promoting growth of bacterial yeast and molds. Where growth has developed significantly flux rates may be restored by removing the growth with flows at high velocities, with the aid of detergents, and by flushing plastic foam balls through the system periodically at intervals of several days to a week or more depending upon the degree of growth experienced. We have been concerned with finding ways to prevent or inhibit such growth, and high velocity appears to be an especially effective method of keeping the system clean or at least reducing the frequency of need for cleanups. Despite the extensive experience with microbiological fouling on sustained runs, we have been unable to observe and to maintain and to analyze fouling rates satisfactorily under carefully controlled conditions ranging to runs of as much as 120 hours of continuous operation with our standard test solutions made up of a calcium-base feed sulfite liquor. Significant increases in the resistance to permeation has been observed at lower velocity with neutral sulfite semichemical white water. These studies are being continued to more closely develop knowledge of the fouling rates and conditions for preventing the development of microbiological fouling.

Microbiological fouling should not be confused with the compaction effect of operating at elevated pressures. Compaction of the porous membrane layer is certainly responsible for the decrease in the membrane constant observed in the early periods of operation at higher levels of pressure within the system. For new modules we have found the flux rate losses range up to 50% in the first 20 hours of operation. This is experienced on all types of substrates in and out of the field of processing industrial waste streams.

CONCLUSIONS

1. A velocity of 1.0 ft/sec is sufficient to produce turbulent flow up to 35.0°C and 4.0% solids concentrations of all four liquors. For higher concentrations a velocity of 1.0 ft/sec may or may not be turbulent depending on the temperatures of the liquors.

2. Pressure drop and pumping energy for the four pulping and bleaching effluents studies are maximum for N.S.S.C. white water and minimum for kraft bleach effluent liquors.

3. Kraft bleach effluent liquors have the highest osmotic pressures which increase linearly from 84.0 psia at 10.0 g/l solids to 594.0 psia at 100.0 g/l solids. The osmotic pressures of the other three liquors range to about 300 psia up to 10.0% solids concentration.

4. Rejection of OD at 281 nm is above 98% for all liquors except for ammonia-base acid sulfite liquor, which averaged 97%. The rejection of solids is above 95.0%, whereas BOD_5 rejections vary between 85-95%. COD rejections range higher than for BOD_5 and for solids rejections.

5. The increase in flux rate per °C rise is 2.1% for calcium-base acid sulfite liquor in the solids concentration range of 1.0-10.0%. The flux rate increase with increase in temperature is higher for water, and is about 2.8% for each °C increase in temperature.

6. Reynolds numbers of the order of 16000-22000 (equivalent to 3 to 5 ft/sec in a tube 0.5-inch inside diameter) may be necessary to prevent concentration polarization and fouling in 12-56 g/l concentrations range of N.S.S.C. white water. For 10% solids calcium-base acid sulfite liquor, the average decrease in flux rate over 70 hours of continuous operation was reduced from 8.0 to 4.0% by increasing the velocity from 0.8 to 1.2 ft/sec. This was true for liquors in which calcium salts had not precipitated. The flux rate decline observed in liquors in which precipitation of calcium salts occurred was observed to be relatively more dependent on the velocity.

Acknowledgment

This project has been authorized, supported and financed in part, by the Department of the Interior pursuant to the Federal Water Pollution Control Act.

REFERENCES

1. Aggarwal, J. P., and Sourirajan, S., Ind. Eng. Chem., 61(11), 62(Nov., 1969).

2. Ammerlaan, A. C. F., and Wiley, A. J., Chem. Eng. Progr. Symp. Series, Water-1969, 65(97), 148(1969).

3. Ammerlaan, A. C. F., Lueck, B. F., and Dubey, G. A., Tappi 52
    (1), 118(Jan., 1969).

4. Kopecek, J., and Sourirajan, S., J. Appl. Polymer Sci., 13,
    637(1969).

5. Vos, K. D., Burris, F. D., and Riley, R. L., J. Appl. Polymer
    Sci., 10, 825(1966).

6. Wiley, A. J., Ammerlaan, A. C. F., and Dubey, G. A., Tappi 50
    (9), 455(Sept., 1967).

7. Wiley, A. J., Dubey, G. A., Holderby, J. M., and Ammerlaan, A.
    C. F., J. Water Pollution Control Federation, Part 2, R279
    (Aug., 1970).

ELECTRODIALYTIC RECOVERY OF SULFURIC ACID AND IRON CONTENT FROM SPENT PICKLING LIQUOR

A. H. Heit

Gamlen Chemical Co.

Dr. C. Calmon

Sybron Corporation

Historically, attempts to apply electrodialysis to the recovery of sulfuric acid from spent pickling liquor go back to the early fiftys, at a time when the first reasonably stable, synthetic ion exchange membranes became available. Thus H. C. Bramer and J. Coull of the Mellon Institute and C. Horner and G. W. Bodamer of Rohm & Haas disclosed within the same year (1955) in Industrial Engineering Chemistry (Vol. 47) two independent investigations of the feasibility of such recovery; the latter pair of researchers were awarded a patent for their particular approach (U.S. Patent 2,810,686(1957) While some differences in flow pattern through these early cells were indicated, each functioned on the electrodialytic displacement of sulfate ion into an anode chamber across an anion permselective membrane, until the concentration of free acid had diminished to a level which permits the cathodic reduction of Fe^{II} to elemental iron and the growth of spongy or dense or dendritic deposits of the metal on the cathode. Both featured dependence on the electrolytic decomposition of water at the anode to supply hydronium ion to complement the incoming sulfate ions, so that in effect the acid is reconstituted in the anode leg of the system.

The problem was reinvestigated by D. J. Lewis and F. L. Tye of Permutit Ltd. (J. Applied Chemistry in 1959, pp 279-292) and later by C. Mantell and L. G. Grenni of the Newark College of Engineering who published in the Journal Water and Waste (34,951-61(1962)

Tye and Mantell assessed the energy requirements,
current utilization and basic cost parameters of the
fundamental two chamber cells with a single anion
exchange membrane serving as a selective barrier
between a cathode and an anode compartment. Mantell
concluded that assuming solubility of the problem of
maintaining free flow through the cathode compartment
due to blockage by electrodeposited iron, the economics
were at best dubious because of low current utilization,
instability of the membrane in highly acidic media and
the doubtful market value of spongy (pyrophoric) iron.
By employing stainless steel cathode, the metal can be
layed down as a spongy deposit, still requiring special
attention to clear the compartment, but nonetheless
having some dollar value which hopefully would work
towards the lessening of operating costs. Mantell
estimated that, if one large steel mill were to be
committed to the process, it would yield more pyrophoric
grade iron than the country sold in 1960-1, would
depress price and hence would confer at best, shaky
by-product credit in the processes economic justification.

In point of fact no answers to the problem of
disengaging the accumulated iron metal in the cathode
without interruption of the process have appeared to
date.

Calmon and Heit have been issued a patent (U.S.
Patent 3,394,068) for an approach to the electrodialytic
treatment of spent pickling liquor, which completely
circumvents this fundamental difficulty of prior art.
Their cell is arranged so that the iron content of the
spent liquor is removed as hydrous oxides of iron which
can be freely moved in a slurry and subjected to by-
pass filtration or centrifugation for separation from
the water phase.

The essential cell system is illustrated in Fig. I.
In common with Bremer, Bodamer et al, the process is
dependent on the electrolytic decomposition of water
with the distinction that both anodic and cathodic
break-up of water are entailed, ie. $2H_2O \xrightarrow{4e} O_2 + 4H +$
$2H_2O \xrightarrow{2e} H_2 + 2OH.$

The cathode chamber furthermore is filled with and
has circulating through it a solution of sodium
hydroxide; no spent pickling liquor is permitted to
enter the cathode region. The mechanism of achieving
sulfate transport away from the spent pickling liquor
is in fact therefore quite unique.

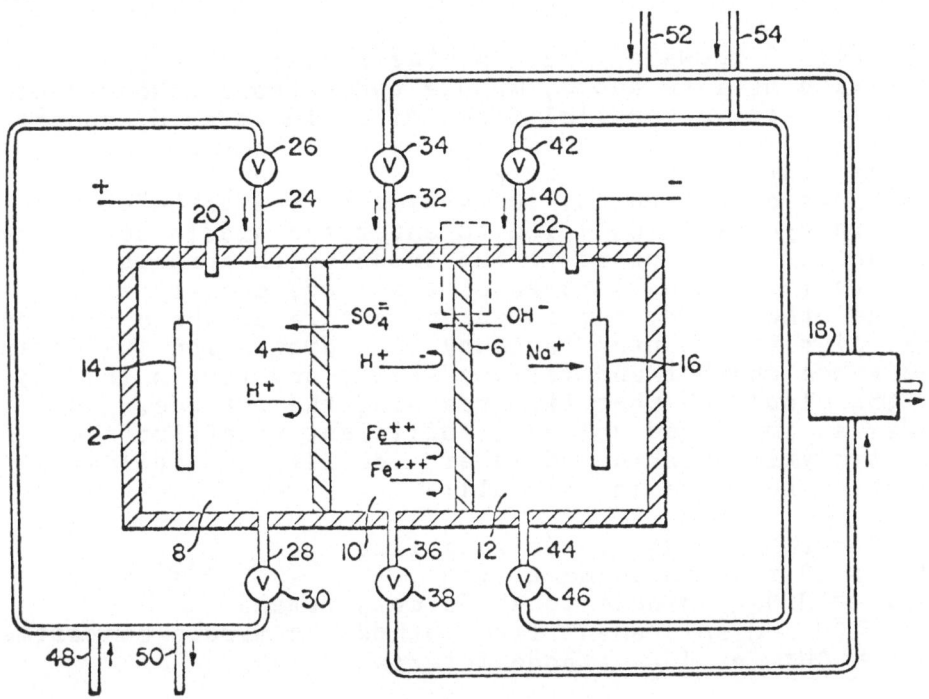

Figure I

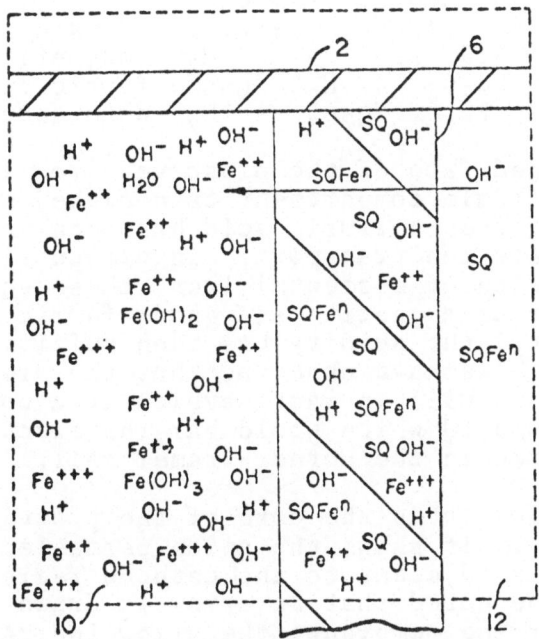

Figure II

Fig. I shows a three chamber system, namely partitioning into anode, middle and cathode compartments with respective parallel flows of sulfuric acid, spent pickling liquor and sodium hydroxide. The middle compartment is defined by a pair of anion permselective membranes one of which, juxtaposed to the cathode, permits hydroxyl (OH^-) ion to enter the middle compartment and another juxtaposed to the anode which permits sulfate ($SO_4^= + HSO_4^-$) to enter the anode compartment. On the other hand, the selective nature of these membranes restrains the counter tendency of hydrogen ion (H_3^+O) in the anode chamber and ferrous (Fe^{++}) and ferric (Fe^{+++}) in the middle chamber from reaching the cathode. Thus under an impressed voltage, sufficient to effect the electrolytic break-up of water, ideally, the following events take place in the cell:

$$2 H_2O = 2H^+ \quad + \quad O_2 \quad \text{anode interface}$$
$$SO_4^= = SO_4^= \quad \text{displacement from}$$
$$HSO_4^- = HSO_4^- \quad \text{middle to anode compartment}$$
$$OH^- = OH^- \quad \text{displacement from cathode to middle compartment}$$
$$2H_2O = OH^- \quad + \quad 4H_2 \quad \text{cathode interface}$$

Middle Compartment

$$(OH^-) \quad + \quad H^+ = H_2O \quad \text{neutralization}$$
$$Fe^{++} \quad + \quad 2 = (OH^-) \quad Fe(OH)_2 \quad \text{ferrous hydroxide}$$
$$Fe(OH)_2 \quad + \quad O_2 = FeO \quad Fe_2 \, O_3 \quad X \quad H_2O \quad \text{magnetite}$$
$$Fe(OH)_2 \quad + \quad O_2 = Fe_2O_3 \quad X \quad H_2O \quad \text{gamma ferric oxide}$$
$$H_2O \quad + \quad Fe(OH)_2 = Fe \, O . Fe_2O_3^2 \quad + \quad H_2 \quad \text{auto decomposition}$$

The indicated fate of the dissolved iron ionic species in the middle compartment take place, when the preponderance of free sulfuric acid has been in effect removed by sulfate ion transport. In common with the cathodic deposition of elemental iron, these equilibria which yield hydrous magnetite and gamma ferric oxide are not evident, until the acidity has been sufficiently lowered, although we have observed that the precipitation of the iron oxides will in fact develop at a concentration of free sulfuric acid which would bar the electro-deposition of iron in the Horner-Bramer cell.

Fig. II illustrates the core of the practicality of this approach. It shows the anion permselective membrane which is adjacent to the cathode region of the cell. It will be noted that an arrow is drawn across the thickness of the membrane; the arrow indicates the direction of transport of hydroxyl ion (OH^-), namely towards the anode. It will be further noted that the illustrated cross section of the matrix of the membrane has scattered through it symbols denoting H^+, Fe^{++}, Fe^{+++}, SQ Fe^n, SQ, OH^-

The symbol SQ refers generically to a sequestrant and is meant to denote a specific type of sequestrant, namely having:
1. Good capacity to deter the precipitation of ferrous and ferric oxides.
2. The fundamental property of a cationic substance ie. the capability when in a region of acidity to migrate back to the cathode compartment by virtue of its positive charge, although it behaves in the sequestrant sense anionically due to reaction with hydroxyl ion.
3. Commercial availability.

In the absence of such a sequestrant, the described electrodialysis is totally impractical by virtue of the rapid fouling of the cited anion exchange membrane (Fig. II) with matrical occlusions of hydrous iron oxides whose presence is registered within the first 30 seconds after voltage is thrown across the cell, by an enormous increase in the ohmic resistance; the current in consequence drops to the micro-ampere range. It renders the whole scheme into an academic exercise, since without the capability of sustaining high current density (eg. 200-500 ampere/ft^2) at relatively low potentials, the energy requirements are insupportably high.

This program has point and meaning in the very pragmatic sense in that it must prove viable for a steel plant with a stream pollution problem due to the disposal of the spent pickling liquor. "Viable" means in turn that the concept must demonstrate that the disposal can be accomplished with recovery of acid and isolation of the dissolved iron on a reasonable economic basis.

Specifically we have found that the sequestrant, triethanolamine works outstandingly well in satisfying the above criteria.

1. It is cationic in nature $N(CH_2. CH_2. OH)_3 + H^+ = (HN(CH_2.CH_2 OH)_3)^+$
2. It also functions in a highly alkaline medium anionically

$$N(CH_2\ CH_2\ OH)_3 + (OH^-) = N\ \begin{array}{c} OH_2.CH_2.O- \\ CH_2.CH_2O- \\ CH_2.CH_2.O- \end{array} + 3H_2O$$

3. It has excellent capacity to sequester both
    $Fe^{++} + Fe^{+++}$

4. It is commercially available at reasonable cost.

Triethanolamine can, in fact, be replaced by
other homologous or cognate substances. We have,
however, executed almost all our bench tests of this
electrodialysis process with it.

While it is relatively low cost material, the
indefinitely continued loss of a material in the
$.30/lb category in the recovery of sulfuric acid at
$.015 to $.02/lb value and iron oxides of no determinate
value, is obviously another insupportable limitation
which again would make the process unfeasible.

The unique aspect, marking the employment of
triethanolamine (TEA) as a membrane anti foulant in
the above sense, is that there is no loss to the
ultimate waste stream, ie. the spent pickling liquor
circulated through the middle compartment: at least
we were unable to detect any sensible loss in any of
the runs with real and simulated spent pickling liquor
over time periods of up to 32 days as long as the
membrane which the TEA is protecting is intact, ie. not
physically damaged due to cracking, fissuring, pin
holing etc. Our technique of measurement of TEA
persistence in the catholyte was primarily based on
periodic differential titration of the catholyte,
noting the inflection point due to sodium hydroxide,
the dominant solute, roughly 2 gm equivalent/lit and
TEA, the minor solute present to the extent of 8 gms/lit
(.0535 molar); in addition Kjedahl nitrogen determinations
were done on the middle compartment liquor intermittently.
From these analyses, we could not spot any significant
loss of TEA.

We attribute this most vitally important conservation
of the critical sequestrant to the ampholytic nature of
the selected material. In the actual operation with
voltage maintained across the system and spent pickling
liquor, (SPL) circulated through the middle compartment,
the face of the membrane (contiguous to the cathode)
which confronts the SPL is almost constantly on the acid
side of the pH fence, initially very low; towards the
termination of the run, when the free sulfuric acid has

been, in effect, majorly removed, at least a pH of 3 to
4 prevails, which is still acidic enough to convert the
TEA to its cationic form and thus under the influence of
the imposed potential, "force it back" electrophoretically
towards the cathode, where it is nonetheless behaving as
a pure anionic entity. The latter behavior is, of
course, quintessential to its function as an anti
foulant, since it must migrate towards the anode,
present itself within the fouling region of the matrix
of the membrane and, through its complexing impact,
overcome the otherwise inescapable precipitation of
intruding iron ions concomitant with the flux of the
principal negative ion ie. the hydroxyl ion (OH).

EXPERIMENTAL DETAILS

A schematic (Fig. III) is attached to this report
to show the relative disposition of 5 gallon polyethylene
storage vessels which served as the reservoirs from which
the three liquors were circulated through the cell's
three chambers.

The cell was electrically connected to a rectifier
rated for an output of 0-200 amperes, 0-30 volts, with a
three phase 220 volt, input (manufactured by the American
Rectifier Corporation, Flushing Point, Queens, New York,
New York).

The cell afforded an active, wetted working, cross
section of 25 square inches (5" X 5") with an overall
dimension of 7-5/8" X 7-5/8". It was fitted with two
specimens of 7-5/8" X 7-5/8" dimensions of Ionac's
anion exchange membranes, coded MA 3475R, which were
especially fabricated in commercial production to with-
stand extremes of acidity and alkalinity.

Some of the characteristics of the Ionac MA 3475R
material are: Areal resistance - 5 ohm-cm^2 in 1.0 N
Na Cl, Mullen Burst - 220 p.s.i., Thickness - 12 mils,
Permselectivity - 99% (0.1 N Na Cl: .05N Na Cl),
Salt Diffusion < 1.0 X 10^{-9} gm equiv/hr/cm^2 (based on
0.1 N Na Cl $=$ H$_2$0). Type - heterogeneous, composed of
a thermoplastic fluocarbon polymer serving as the binder
for a strong base, di vinyl benzene cross-linked
polystyrene anion exchanger, supported on woven polypropylene web. Ion Exchange capacity: 1.1 meq/gm
(anhydrous basis). Gel Water - 18% (after 24 hrs
immersion in water. The membranes are originally
anhydrous)

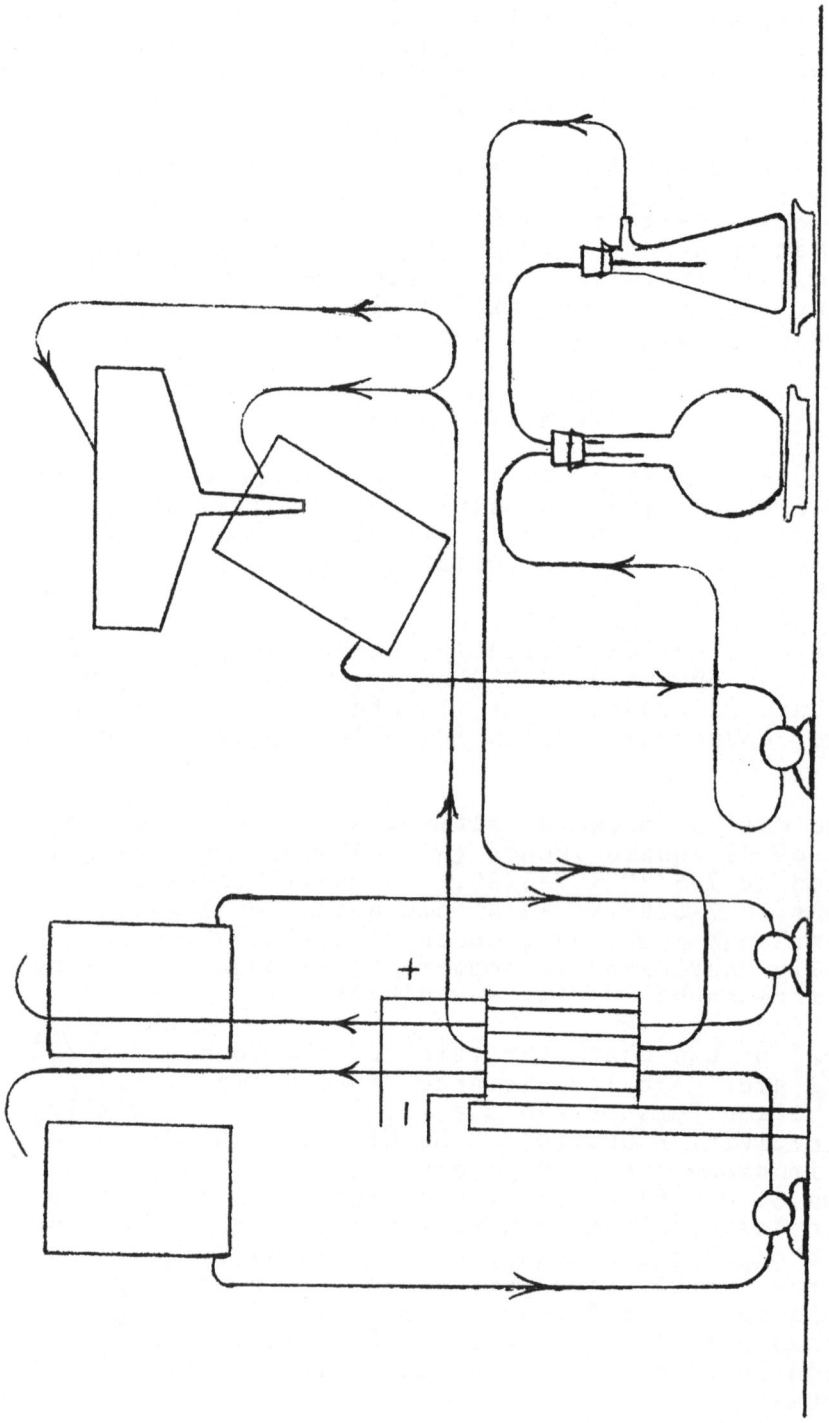

Figure III

The frames were cut from 1/2" thick rigid PVC sheet, bored at the corners along the diagonal and the borings fitted with 2" PVC nipples which were solvent welded to the frame. Due to the cold flow property of PVC, operation was confined to 50-55^{oc}(max.)

Gaskets as well as feed lines were of Tygon. The former are approximately 1/16" thick (approximately), since some diminution of thickness ensues on closure of the cell with a group of six large "C" clamps, distributed at each corner and at the middle of the top and bottom of the cell pack.

We employed the following quantities of solutions:

1. Catholyte 16 liters containing 640 gms. sodium hydroxide (reagent grade, pellets), 130 gms. TEA (practical grade, Matheson, Coleman & Bell) made up to volume with demineralized water.
2. Eighteen liters of spent pickling liquor from the Bethlehem's Lackawanna New York Plant containing 26.9% w/w H_2SO_4, 7.9% w/w $FeSO_4$. Density = 1.23 at 70°F. Samples during the run were analyzed for Fe^{++}, Fe^{+++} and for H_2SO_4. The total mass of Fe^{++} and Fe^{+++} remained constant until the H_2SO_4 content had declined to 60.9 gms/liter.
3. The anolyte initially consisted of a solution of 15% w/w H_2SO_4 made up by dilution of 1.356 liters of reagent grade sulfuric acid (Sp. Gr. 1.84-96-98% Assay) in 13.504 liters of demineralized water affording a total volume at ambient temperature (23-25^{oc}) of 14.1 liters. Samples were titrated against 1 N NaOH to phenolphthalein ED to fix the increase of its H_2SO_4 content in the operation of the cell.

Each of the three liquors were circulated upflow through the three chambers of the cell. In the instance of the catholyte, a sealess magnetic drive turbine pump with a polypropylene body and impeller was used (Cole-Parmer Model 7004-3, 1/35 HP, 115 volt, 350 g.p.h. at 3 ft. head, 330 g.p.h. at 6 ft. head). This pump functioned without incident throughout the 32 days of the run with the Lackawanna liquor as well as approximately 60 days of antecedent studies involving simulated spent liquor (16-17% $FeSO_4$, 6-7% H_2SO_4).

In the instance of the middle chamber, a pump of different construction was needed in view of the ultimate formation of a slurry of magnetic iron oxide particles as well as the corrosive nature of the spent liquor at the

outset of a run; the eventual appearance of hydrous
magnetite and gamma ferric oxide would soon clog a pump
featuring the magnetic drive principle. We chose a
pump manufactured by the Vanton Pump Co., Irvington,
New Jersey, who also offer a sealess pump which is
trademarked "Flex-I-Liner", features a Teflon body
block and Kel-F Elastomer liner, 1/4 HP, 1725 RPM,
enclosed fan cooled motor, 115 volt, 60 cycle A.C. 1
gal/min, maximum discharge pressure 25 p.s.i. Cole-
Parmer Model 7073T. This also performed smoothly in a
period of over three months of more or less continuous
24 hr./day use.

As illustrated in the flow sheet (Fig. III), the
middle compartment liquor is the spent pickling liquor
and has a more complex path between reservoir and the
chamber through which it is recirculated. The extra
features were decided upon in anticipation of the
handling of the eventual development of variously
particle sized iron oxides and the necessity to lower
the concentration of suspended particles in order to
avoid the clogging of the middle chamber with sediment.
Thus, we have -

1. Tilted the middle stream reservoir to obtain a
    region for oxide sedimentation.
2. Interposed two additional sedimentation vessels, a
    6 liter florence-flask and a 2 liter side-arm,
    suction flask, both mounted on magnetic stirrer
    plates, which, when the precipitation of the oxides
    was reached, were plugged to the house circuit,
    providing a rotating magnetic field which crudely
    simulated the cyclonic separators widely used in the
    paper pulp industry to effect the removal of iron and
    iron oxide fragments from the fibre-slurry before
    commitment of same to the Fourdrinier or Cylinder
    machine.
3. Mounted a large (16 lit.) Buchner funnel over the
    mouth of the middle compartment reservoir. The
    funnel was lined with polypropylene felt and
    connected through bypass lines with the return lines
    of this circuit so that when massive formation of
    iron oxide particulates developed, complete removal
    from the system was possible by gravity based
    filtration.

Of course, the cell frames could have been constructed
of polypropylene, which can endure temperatures up to
90-95°C without the cold flow complication. In point of

fact, a cell configuration, experimented with before the arrival of the Bethlehem liquor, was made of polypropylene. The latter was of a pentagonal cross-section. With this cell, we were able to push current density up to 750 ampere/ft^2.

CORROSION RATE OF LEAD ANODE

16.4 = Wt. of $PbSO_4$, converted from PbO_2 collected from anolyte in the period 12/2 - 22/68 by these reactions:

$$PbO_2 + 4 HCl \rightleftharpoons PbCl + Cl_2 + H_2O$$
$$PbCl_2 + H_2SO_4 \rightleftharpoons PbSO_4 + 2 HCl$$

$$\frac{16.4}{303} \times 207 = 11.2 \text{ gms. Pb}$$

$$\frac{11.2}{11.34} \frac{in.^2}{16.4 \text{ cm}^3} \times \frac{1000 \text{ mils}}{25 \text{ sq. in.}} \times \frac{365 \text{ days}}{20 \text{ days yr.}} = 44.4 \text{ M.P.Y.}$$

We believe this represents a conservative and therefore a high figure for the corrosion. The rate of sluffing of the lead dioxide from the lead anode is actually maximal at the outset of the run, while it does continue through the run, it never reaches as high a rate as at the beginning when it is first forming from the virgin lead. Once a coherent deposit has developed, its rate of attrition is substantially lower. Our value reflects the accumulation in the anolyte from the inception of the run to the twentieth day.

At this rate of penetration, a 1/2" thick lead plate will lose in 365 days $\frac{.044}{.500} \times 100 = 8.8\%$ of its thickness. If a single such sheet of lead is used as the basis of serving two contiguous three chambered cells, then the maximum loss of thickness should amount to 8.8 X 2 = 17.6%, a safe enough decline which still permits the sheet to hold to its hydraulic integrity and rigidity.

On the general subject of corrosion, we had earlier ascertained that the following materials for anode construction are grossly unsuitable.
(a) Platinized titanium; a specimen with a coating of 125 micro inches of platinum ($63.00/ft^2) on 40 mil thick titanium was completely deplatinized within 48 hours under comparable conditions.

(b)   Graphite, of the type used in the Cl_2 - NaOH
diaphragm cell, was completely perforated within 48 hours
of operation at 288 amp./ft^2.

   In this same vein of the critique of materials of
construction, we have determined that copper is
unsuitable for the cathode. While copper sheet was not
perforated, significant penetration was observed within
a 3 day period, so that a palpable ridge line defining
the active electrode area was observed.

   The Ionac membranes were found to have excellent
stability, provided that antibillowing, perforated
rigid PVC sheet was interposed between the membranes and
the adjacent chambers. Without the buffering effect of
these perforated plates to restrain mechanical distortion
of the membranes, damage did occur, particularly where
the membrane physically impinged on the electrodes. The
latter complication was due to the unequal pressure
delivered by the pumps, the middle compartment being
marked by a greater pressure than either adjoining
compartment.

   FORMATION AND ISOLATION OF MAGNETIC IRON OXIDES

   The system behaved well during the course of the
electrodialytic displacement of sulfate ion associated
with the free sulfuric acid content of the spent
pickling liquor. There was a gradual increase in the
voltage drop across the electrode as the concentration of
sulfuric acid declined in the middle compartment reaching
11.3 volts at 8:55 A.M., 12/22/68, when the sulfuric acid
in this section had dropped to 1.1 Normal (60.4 gms.
H_2SO_4/lit. from an initial 330 gms/lit.) The original
voltage was 8.0 at outset, 12/2/68 which held until
12/22/68.

   By the 25th of December, with the first manifestation
of oxide solids (at pH = 0.7), the voltage drop rose
sharply; 12.9 volts delivered only 15 amperes across the
stack. Adjustment of the rectifier restored the current
to 25 amperes at a voltage of 27.8 volts. On the 26th,
there was another decline in amperage which reached
10-12 amperes. It was noticed, however, during this
period that by manually pulsing the feed lines to the
middle compartment, sharp increase in amperage followed.
The effects, however, were not long lasting and the
current within a half hour subsided back to the 10-12

ampere level. It was suspected that a scale of iron oxide was forming at the middle compartment of the anion exchange membrane juxtaposed to the cathode. Flexing the feed lines resulted in the mechanical disengagement of the oxide scale from this face of the membrane, but resumption of the smooth flow of liquor permitted the continued growth of the oxide-crystals. The fact that a relatively still substantial amperage prevailed even with this growth of oxide crystals indicates that the critical matrix of the membrane remained unfouled.

In summary, the system was behaving poorly at the juncture of oxide formation. It was then realized that, if a means could be developed which would obviate smooth flow past this membrane, the scale could be continuously dislodged. Because of the higher voltage, the temperature of the system had risen, and a fair rate of evaporation marked the passage of the middle compartment liquor by-passed through a large Buchner funnel mounted into the neck of the reservoir lined with a matte of poly-propylene fibre which supplemented the sedimentation effects in removing suspended iron oxides.

We decided not to replace this water lost due to evaporation and determine what would happen, if in the return from the middle compartment reservoir, air would be sucked into the stream through the drainage nipple at the base of the tilted reservoir, once the level of this liquor had been sufficiently lowered. We speculated that the air pulled into the feed line to the pump and thence to the cell would yield intermittent surges of line pressure and thus assist in the disengagement of the scale.

This is precisely what happened. No further effort was needed to manipulate the feed lines to the middle compartment. In point of fact, once a dependable rate of air bleed to this leg of the system was established, the current rose to as high as 75-80 amperes (432-464 amp/ft^2) at 27 volts, which obliged us to lower the voltage to 16.8, whereupon a steady 50-55 ampere was maintained with the temperature ranging to 40° in the catholyte and middle compartment and 55° in the anolyte. The precipitation of iron oxides during this period was so heavy and voluminous that the rate of return of filtrate from the Buchner funnel was too slow to ensure an adequate supply of liquid to sustain the pumping of the middle compartment liquor. A technique then evolved, consisting in the intermittent use of the by-pass feed

line to the Buchner and the intermittent stopping of
such flow to the funnel, permitting the filtrate to
return to the reservoir and thus sustain a working level
therein. The rapidity of oxide formation was so great
that we were also obliged to remove the oxide sludge
from the Buchner funnel within an hour or two of its
initial bulk appearance at this higher current density
(288-300 amp/ft^2). Because so much of the liquor was
entrapped in the sludge, we were also obliged to make
up a modest additional quantity of ferrous sulfate
solution and add this to the middle compartment. This
last comprised 500 gms. $FeSO_4.7H_2O$ in 3500 ml. water
(yielding a volume of 3.75 liters).

The sludge, per se, was scooped out of the large
by-pass Buchner and transferred to another Buchner where
it was pulled to a dense cake, washed with water and
subsequently dried in a glass tray at about 50-60^{OC}.

The run was finally ended on January 2, 1969, some
768 hours from its inception.

In summary, feasibility has been established for a
method featuring the employment of anion exchange
membranes in a simple three chamber electrodialytic cell
to effect the recovery of essentially iron free sulfuric
acid and the isolation of the iron value from a spent
pickling liquor as magnetic iron oxides. The claim of
feasibility is based on the following: (1) The
exhibition of chemical and mechanical stability of the
membranes in a continuously operating treatment of a
spent liquor of abnormally high free sulfuric acid
content, terminated at the end of 32 days (768 hours).
(2) The stability of lead as a material of anode
construction. (3) The achievement of concentration of
sulfuric acid in the receiving anode leg of the system
as high as 28.5% from an initial concentration of 15%.
(4) The ability to sustain a current density of 144 ampere/
ft^2 with no evidence of membrane fouling due to iron
oxide deposition or organic matter present in the un-
filtered specimen of spent liquor obtained from the
Lackawanna, New York Plant of the Bethlehem Steel Corp.
(5) The indications of acceptable energy requirement on
operation at elevated temperatures, namely, 80-85^{OC}, a
projected rating of 1.5-2.0 KWH per pound of 100% H_2SO_4.
(6) The reduction of the electrolytic content of the
spent pickling liquor from 26.8% sulfuric to 0.00%
H_2SO_4 and the ferrous sulfate content from 7.9% to
0.00%. (7) The isolation of the iron values as exclusively

magnetic iron oxide. (8) The freedom from metallic
iron deposition on the cathode. (9) The retention in
the catholyte of the original quantity of triethanolamine,
essential to the uninhibited transport of hydroxyl ion
into the middle chamber, thus establishing that the issue
of make-up of this costly additive is minor, if not
negligible. (10) The retention in the catholyte of the
original quantity of sodium hydroxide, its principal
electrolyte, also establishing the freedom from any
necessity of make-up. (11) The purity of recovered
sulfuric acid, less than 200 ppm contamination by ferric
ion. (12) The constancy of the slope of the plot of
productivity (lbs 100% H_2SO_4/ft^2/day) against current
density (ampere/ft^2) a constancy - $\underline{.0333 \text{ lbs } 100\% \text{ } H_2SO_4}$
                                           ampere day
which enables dependable calculation of productivity in
the range 100 to 800 ampere/ft^2 and therefore design of
a system to cope with a given daily output of SPL.

The attached tables are intended to provide:

1. An estimate of the capital investment and operating
   costs in applying this particular process to a steel
   mill wherein 200,000 tons/year of steel is pickled
   with sulfuric acid (cf. Tables I & II).
2. A realistic overview of the prevailing practice of
   pickling in the steel industry. The severe decline
   in the employment of sulfuric acid and the sharp
   ascent in the tonnage treated with hydrochloric acid
   is to be noted (cf. Tables III & IV).

Regarding the possible application of the above
described electrodialytic process to recovery of
hydrochloric acid, the authors of this paper believe
that, by a relatively minor modification of the cell
structure and flow sheet, it can be adapted. The sole
change consists in the interposition of an additional
compartment between the middle and the anode compartment.
This extra compartment is defined by a cation exchange
membrane, juxtaposed to the anode section; such a
membrane would, effectively exclude chloride ion from
reaching the anode, provided the hydrochloric acid
concentration were limited to a maximum between 4 and
5%. The original three chambered set up is inapplicable
by reason of the ease of anodic oxidation of chloride
ion to elemental chlorine, chloric, perchloric acid,
the consequent loss of viable hydrochloric acid and the
breeding of a host of most difficult corrosion and
handling problems.

TABLE I

ELECTRODIALYSIS - COST CALCULATION

Plant - Steel Production Pickled	200,000 tons/year
Capital Cost	$597,000
Rectifier & Elect. Equipment	277,000
Membranes Stock	112,000
Operating Cost	97,100/yr.
Volume of Waste	1.65×10^6 gals/yr
Conc.	Up to 2.7×10^6 gal
$FeSO_4$	330 gms/lit
H_2SO_4 (free & combined)	210 gms/lit
H_2SO_4 to be conc. to	30%
Tons of H_2SO_4 to be recovered	415 tons/yr
Loss of H_2SO_4	451 tons/yr
Rectifiers	
Amperage	876,000 amps.
Voltage	3.1 vols
Oper. Cost/gallon	3.55¢/gal.
Dumping method for the above	
Plant-Capital	$125,000
Operating Cost	
(Dumping 8,000	
Chelate 4,000	
Acid Value 59,000)	
Operating Cost per gallon	2.75¢/gal.

TABLE II

OPERATING COSTS OF E.D. PLANT

Total Cost	$96,000
Acid Make-Up	14,000
Chelate Additive	4,000
Power Cost (1.9 KWH/lb. H_2SO_4)	33,000
Labor	15,000
Maintenance	27,000
Rectifiers(60%)	
Membranes(13%)	
& Misc.(27%)	3,000
Value of Acid Recovered @ 31.69/ton 66°Be	
No Credit for the magnetite	
Cost/Gallon(2.7×10^6)	3.55¢/gal.

TABLE III

SULFURIC ACID PICKLING OF STEEL PRODUCTS

1.	Total steel products production	130×10^6 tons
2.	Quantity of steel products pickled	40%
3.	Peak volume of sulfuric acid used (1964)	1.1×10^6 tons
	(1970)	4.4×10^5 tons
4.	Strength of acid used	20-25%
5.	Sulfuric acid used	45 lbs/ton (some=30#)
6.	Volume of waste liquor	30 gals/ton
7.	Total waste volume at peak	1×10^9 gals
8.	Residual acid in strong waste liquor	25% of total
9.	Residual acid in rinse water	15% of total
10.	Acid converted to ferrous sulfate	60% of total
11.	Ferrous sulfate in rinse water	15% of total purch.
12.	Ferrous sulfate conc. in strong liquor waste	7-15%
13.	Acid strength in strong liquor waste	7%
14.	Total waste acid at peak	2.5×10^5 tons/yr.
15.	Rinse water required	80 gals/ton
16.	Ratio of rinse to strong waste solution	2.6
17.	Drop of sulfuric acid use from 1964 to 1970	60%

TABLE IV

HYDROCHLORIC ACID PICKLING OF STEEL PRODUCTS

		1963	1965	1969
1.	Total steel pickled with acid-tons	47×10^6.	48×10^6	50×10^6
2.	Steel processed with HCl large plants using HCl	25,000	150,000	30×10^6 80%
3.	Strength of acid used		20°Be	
4.	Acid used		40 lbs/ton	
5.	If HCl recovered, acid required		8 lbs/ton	
6.	Volume of waste liquor strong		8 gals/ton	
7.	Residual acid in strong waste liq.		0.5% of total	
8.	Residual acid in rinse water		15% of total	
9.	Ferrous chloride conc. in strong liquor		27%	
10.	Ferrous chloride in rinse water		15% of total produced	
11.	Rinse water required		80 gals/ton	

TABLE V

POTENTIAL DAMAGE DUE TO WASTE LIQUOR

1. To aquatic life if pH 6.0 and buffered of water reduced.
2. Water can become highly corrosive if used for cooling.
3. Water will require more chemicals for treatment.
4. Can cause line clogging if precipitates form.
5. Oxygen depletion due to iron oxidation.
6. Economic loss of acid and iron.

In this last connection, we have demonstrated, at the bench, this principle in an analogous system of electrodialysis namely the metathesis of potassium chloride and phosphoric acid for the purpose of producing dipotassium phosphate (precursor to the manufacture of the pyrophosphate). Furthermore, C. R. Murphy (U.S. patent 2,865,822) has described the same principle in the purification of pentaerythritol in which a separate stream of formic acid is developed by the transport across a cation exchange membrane by hydronium ion to complement formate ion displaced electrodialytically across an anion exchange membrane which defines one wall of a pentaerythritol sodium formate compartment.

Table V summarizes some of the specific pollution effects which follow the disposal, without treatment, of spent pickling liquor.

SELECTION OF APPROPRIATE RESINS FOR PIEZODIALYSIS

F. B. Leitz and W. A. McRae

Ionics, Incorporated, Watertown, Massachusetts

The very considerable developments of Staverman,[1] Kedem and Katchalsky[2] and others[3] in the application of irreversible thermodynamics to ion-exchange membranes strongly suggest that it should be possible to concentrate a salt solution by passing a portion of the solution through a specially designed membrane. This phenomenon, called piezodialysis, has been observed with sodium chloride solutions.

Piezodialysis is a promising new process for water desalination. In this process a hydrostatic pressure head forces a solution of increased salinity through a specially designed membrane. This process depends upon two factors: that the internal concentration in a modern ion-exchange resin (2 to 10N) is much greater than the salt concentration in most naturally occurring saline waters, and that substantial frictional interaction occurs between the mobile ions and water in such membranes. One type of membrane which may give this performance is a mosaic of discrete areas of anion and cation selective resin each area passing from one face of the composite membrane to the other. A mathematical model[4] has been developed whose predictions closely match the data. Some examples are given below.

This paper focuses on the properties of the resin which are required for piezodialysis. Transfer of cations and anions along separate paths requires two different ion-transfer materials. When the emphasis is on the properties of the individual materials these are referred to as "resins". The composite is referred to as a "membrane".

The mathematical analysis is based on the phenomenological equations for coupled flow.

In this analysis it is assumed that the resins are 100% selective so no coion penetration occurs, that the differences in the various potentials between the bulk liquid phases behave as if they were applied evenly across the membrane phase, that the processes occur sufficiently slowly that irreversible thermodynamics can be applied and that flux coefficients determined with a monofunctional resin, can be applied to segments in a mosaic membrane.

$$J_1 = L_{11}F_1 + L_{12}F_2$$

$$J_2 = L_{21}F_1 + L_{22}F_2$$

The Onsager relationship states:

$$L_{12} = L_{21}$$

In these equations J_i is the flux of species i in gram formula weights (gfw) per cm^2/sec, F_j is the total thermodynamic force on species j in joules/gfw and L_{ij} relates the flux of component i to the force applied to component j in appropriate units. Subscript 1 refers to water and subscript 2 refers to the counterion. With no coion penetration each flux coefficient has a single limiting value. The total thermodynamic force for this process is given by:

$$F_j = \Delta \mu_j + V_j \Delta P + z_j \Delta E$$

The factors μ, P and E represent the concentration dependent portion of the chemical potential, hydrostatic head and electrical potential respectively. Other types of potential (temperature, gravity, etc.) are ignored. Partial formular volume is V_j, and $\Delta P = P' - P''$ where (') designates the high pressure value and (") represents the low pressure value. Since thermodynamics requires that L_{11} and L_{22} be positive, and apparently L_{22} is positive in this process, a flux is positive when it is from the high pressure to the low pressure side.

From this model, equations have been derived for relating permeate concentration and flux to high pressure solution concentration, pressure head and circulating current flow path length for a given set of transport coefficients.

The factor L_{22} is the membrane conductivity. The apparatus used for resistance measurement is shown in Figure 1. This is an adaptation of the device used by Wilson et al.[5] It has the parti-

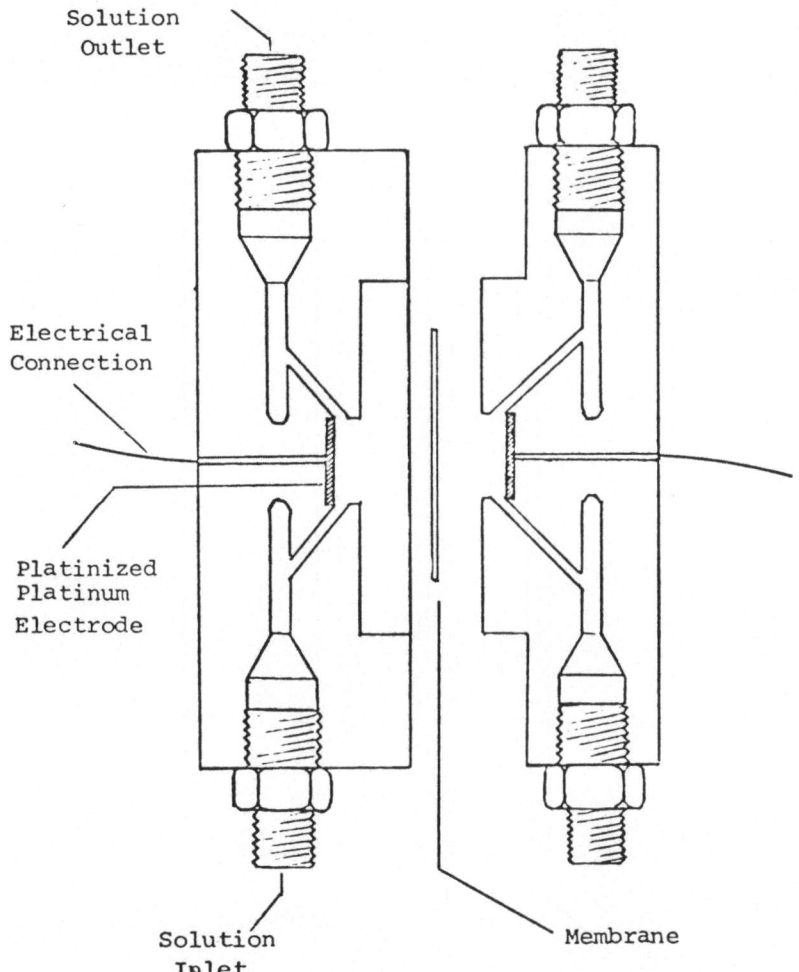

Solution
Outlet

Electrical
Connection

Platinized
Platinum
Electrode

Solution
Inlet

Membrane

FIGURE 1.   RESISTANCE MEASUREMENT CELL

cular advantage that the membrane is in intimate contact with a
solution of known composition. This has been modified from the
original apparatus in that the electrodes have been moved very
close to the membrane surface. This modification was to reduce
the solution resistance which is in series with the membrane resistance. Even with this modification the measurement of a highly
conductive 1 mil membrane may contain an uncertainty of 25%.

The ratio L_{12}/L_{22} is the mole ratio of water to ions transported in the water transport test. The volume method is used
for reasons adduced by Lakhshminarayanaiah.[6] A relatively small

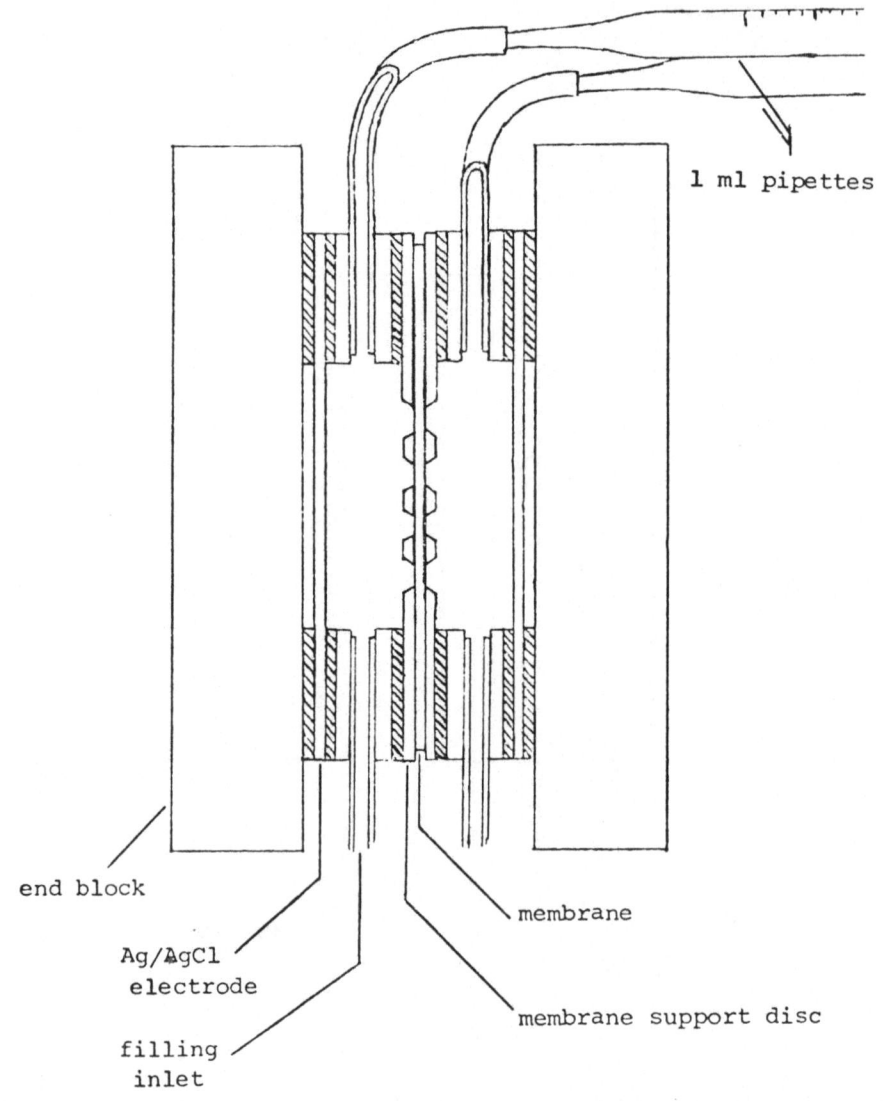

FIGURE 2.   WATER TRANSPORT CELL

volume of solution is used to provide a simultaneous measure of
the degree of coion exclusion. The apparatus is shown in Figure
2.

The third coefficient, L_{11}, is extracted from the water permeation rate with some compensation made for the salt which is
carried through simultaneously.

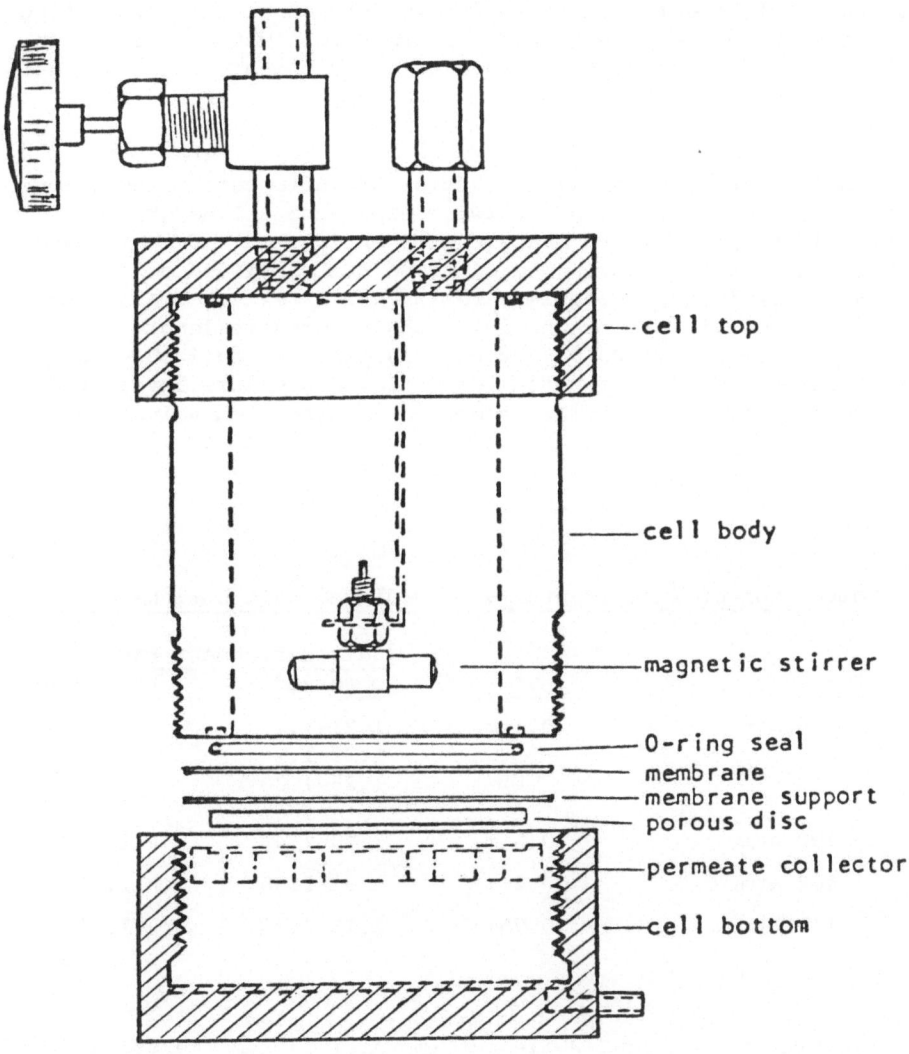

FIGURE 3.   HIGH PRESSURE TEST CELL

$$L_{11} = \frac{J_1}{F_1} - \frac{L_{12}}{L_{22}} \left(\frac{J_2}{F_1} - L_{12} \right)$$

The apparatus for this test is shown in Figure 3. The same test cell is used for testing composite membranes for salt enrichment. In regard to composite membrane tests, some comments should be made about the cell. Early tests were run with a piece of porous

PVC used as the membrane support. Under high pressure this collapsed sufficiently to raise the resistance of the circulating current path to the extent that salt rejection was observed with a membrane which subsequently gave salt enrichment. This situation was rectified by changing the membrane support to an ordinary filter paper.

Having measured the flux coefficients of a cation and anion resin and somehow combined the two resins into a fine grained mosaic, it is possible to measure salt enrichment and to compare it with predicted values. The membrane CR1 consisted of commercial styrenized polyethylene resins which were thermally laminated into a mosaic pattern having a characteristic length of 0.0025 in. (.0064 cm) which, by calculation, is negligible for these resins. Data from batch tests at two pressure levels and three feed concentrations are compared to predicted values in Table 1.

TABLE 1

Comparison of Experimental and Predicted Salt Enrichment

Pressure	C'	Salt Enrichment Factor, α	
		Observed	Predicted
68 atm	0.1 m	1.21-1.27	1.24
		1.22-1.27	
		1.23-1.30	
102 atm	0.1 m	1.40-1.42	1.38
102 atm	0.2 m	1.33-1.36	1.35
102 atm	0.75m	1.16-1.19	1.17

The experimental values, while not impressive from the standpoint of desalination, are very good confirmation of the theory.

Higher values of salt enrichment have been obtained from latex-polyelectrolyte membranes. These membranes are made by mixing a small quantity of sodium polystyrene sulfonate with an emulsion of styrene butadiene copolymer, and reacting a cast film of this material with anion-exchange forming reagents. Salt enrichments for some of these membranes are given in Table 2. We do not have the theoretical apparatus to determine the flux coefficients for the cation resins in these membranes so there are no predicted values for comparison.

TABLE 2

Performance of Some Latex Polyelectrolyte Membranes

Membrane Design- ation	Thick- ness mils	Applied Pressure atm	High Pressure Solution Molality-C'	Solution Flux GFD	Salt Enrichment Factor, α
EM-1b	5.0	68	0.097	5.77	1.35
EM-1c	5.0	115	0.0274	8.48	1.79
EM-1c	5.0	115	0.00244	8.2	5.40
EM-3aX	3.0	115	0.0023	10.75	3.94

In the piezo process salt enrichment is reduced as the external circulating current path length is increased. To approximate the electrical field adjacent to the mosaic membrane, a solution was obtained for Laplace's equation under appropriate constraints.[7] This permitted calculation of the ratio of effective path length to characteristic length, measured across the membrane surface. Runs were made with striped membranes of various stripe widths. In Table 3 the path lengths deduced from salt enrichment measurements (labeled "data") are compared to the effective path lengths calculated from the Laplace's equation solution (labeled "theory"). Reasonable agreement was obtained.

TABLE 3

Comparison of Path Length and Ratio Calculated from Data and Theory

Characteristic Length cm	Path Length cm		Ratio of Path Length to Characteristic Length	
	Data	Theory	Data	Theory
0.0064	~	0.062	~	9.78
0.050	0.6	0.49	12.0	9.78
0.25	4.1	2.53	16.4	10.15
0.55	9.6	7.37	17.5	13.4
1.9	~	67.6	~	35.6

Concentration efficiency is defined as the ratio of reversible pressure required for given high and low pressure solution concentrations to the applied pressure for those concentrations. A value of 20% appears attainable. A reasonable estimate of unit performance is obtained by assuming that the concentration efficiency is constant along the flow path or that an average value of efficiency can be used.

Drawing upon this analysis, which has given predictions in close agreement with experimental results, we expect a strong

influence of the flux coefficients on piezodialysis performance.
We defined a coupling ratio, b, as:

$$b = \frac{L_{12}^{2}}{L_{11}L_{22}}$$

This has a thermodynamic upper limit of unity. Since the coefficients L_{11} and L_{22} are positive definite, the lower limit of this
ratio is zero. This ratio is the most important single criterion
for selection of resins for piezodialysis membranes.

The influence of the coupling ratio on the performance of a
piezodialysis plant is shown below. For simplicity we assume that
the flux coefficients for the two resins are identical.

Unit performance predictions are presented in Figure 4 and
Figure 5. As a simplification sea water was assumed to be 0.6 m
NaCl, concentrated brackish water was 0.06 m, dilute brackish
water was 0.03 m NaCl and the product in each case was 0.006 m NaCl
(350 ppm). The figures present the fractional recovery as product
and the ratio of production rate to permeation rate as a function
of reversible pressure. Since the reversible pressure equals the
product of applied pressure times point efficiency, the abscissas
are so labeled. An abscissa value of 300 or more would result in
very attractive performance. This would result from an efficiency
of 20% and an applied pressure of 1500 psi. These figures strongly suggest that piezodialysis will be run at as high a pressure as
is practical.

Determination of the highest obtainable point efficiency can
be related to the more empirical problem of how great a coupling
factor can be realized. By varying b for given values of other
parameters, the relation between b and peak efficiency can be deduced. One such relation is shown in Figure 6.

Our present means of search for appropriate resins is to
fabricate sheets of candidate resins and test them for coupling
ratio by the techniques described above. No correlation with
properties like water content or capacity has emerged from the
data taken thus far. While we do not expect to formulate general
rules for widely varying types of resin, we hope to develop some
specific rules through a concentrated study on a particular type
of resin. The system under most intense study is aminated PVC.
Quite a range of properties has been obtained by varying the
aminating and quaternizing reagents and reaction conditions. These
have included one resin with a coupling factor of 0.7 . From this
program some understanding of the relationship between membrane
morphology and coupling factor is expected.

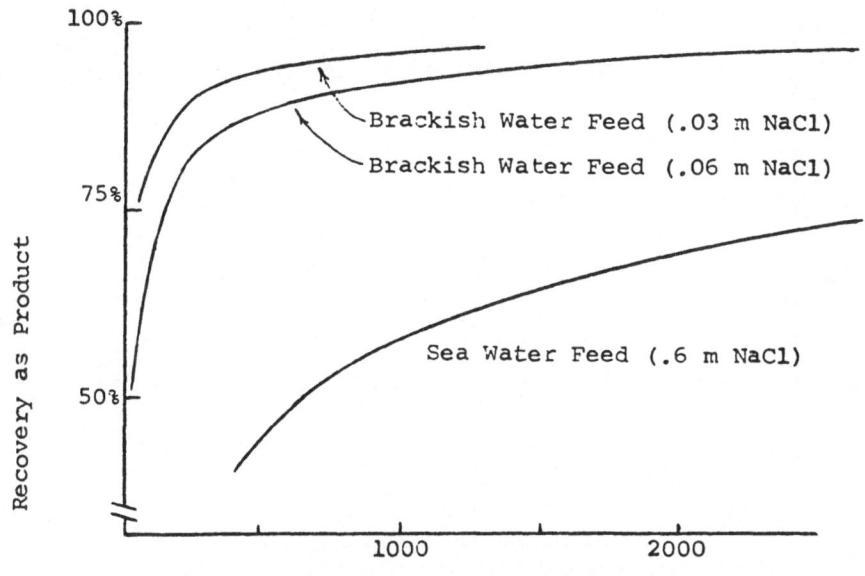

FIGURE 4.   RECOVERY AS 0.006 m (350 ppm) PRODUCT

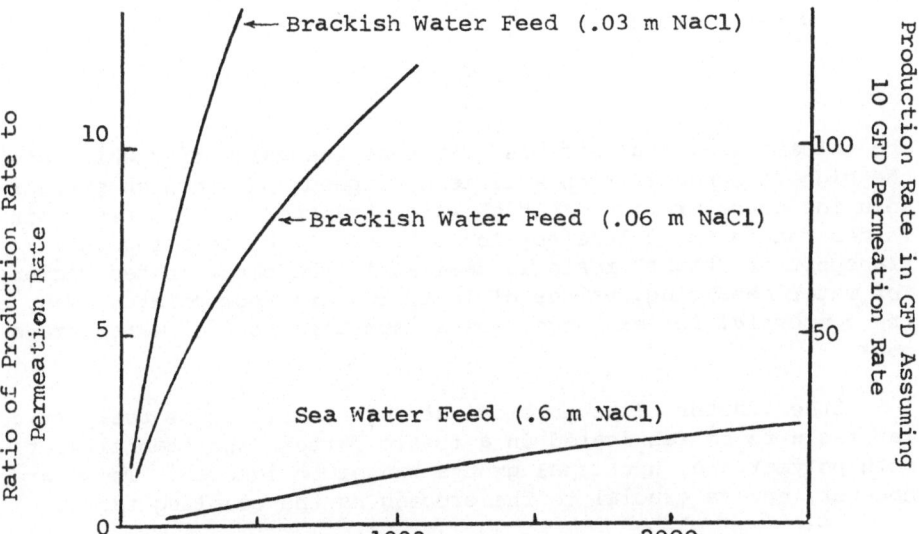

FIGURE 5.   RATIO OF PRODUCTION RATE TO PERMEATION
RATE FOR   0.006 m (350 ppm) PRODUCT

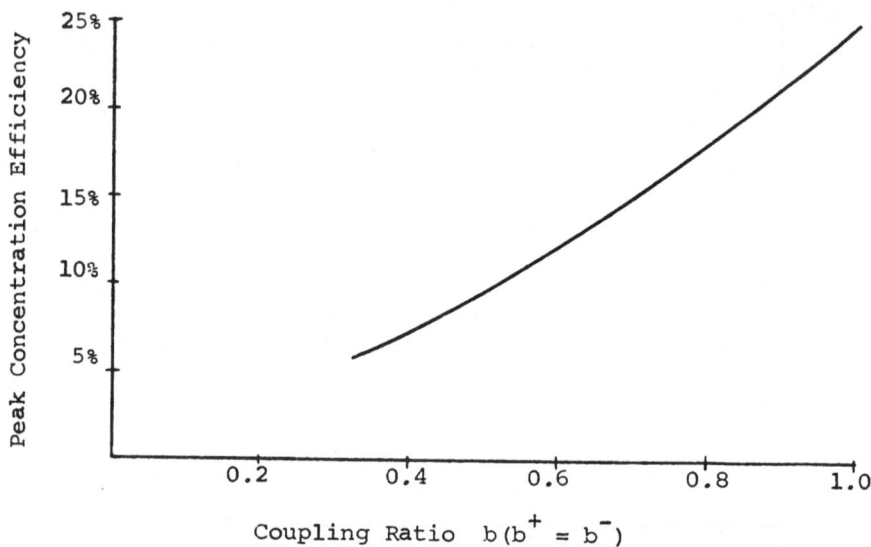

FIGURE 6.   RELATION OF PEAK CONCENTRATION EFFICIENCY
            TO COUPLING RATIO FOR $f^+=f^- = 50, \ell = 0,$
            $\Delta P = 1500$ psi

A water transport ratio, f, is defined as:

$$f = \frac{L_{12}}{L_{22}}$$

The mathematical analysis predicts that the value of f which gives
the highest concentration efficiency depends on the high pressure
solution concentration. A high water transport or "loose" resin
is desired in the dilute portion of the cell while a low water
transport or "tight" resin is desired in the concentrated regions.
For water desalting, values of 10 to 100 are appropriate. Many of
the commercial ion-exchange resins have this sort of water transport.

Other factors like physical strength, ability of the particular resin to be fabricated in a mosaic pattern and stability of
both polymer and functional groups cannot be ignored. Those are,
however, not as crucial to the process as the coupling factor.

This work has been supported by the Office of Saline Water,
U.S. Department of Interior under contracts 14-01-0001-611 and
-2333.

REFERENCES

1.   Staverman, A. J., Trans. Faraday Soc. <u>48</u>, 176 (1952).

2.   Kedem, O. and A. Katchalsky, Trans. Faraday Soc. <u>59</u>, 1931 (1963).

3.   Caplan, S. R. <u>et al</u>, O.S.W. Report No. 413 and Reports on Contract #14-01-0001-2148, U.S. Dept. of Interior.

4.   Leitz, F.B., Proc. 1970 Summer Computer Simulation Conference, Denver, Colorado, 353 (1970).

5.   Wilson, J.R., "Demineralization by Electrodialysis", Butterworth London (1960), p. 211.

6.   Lakshminarayanaiah, N., J. Electrochem.Soc., <u>116</u>, 338 (1969).

7.   Leitz, F.B., C.W.Plummer, J.Shorr and D.M.de Winter, Quarterly Report No. 2, Contract #14-01-0001-2333, Office of Saline Water, U.S. Department of the Interior.

SEPARATION OF HYDROCARBONS BY LIQUID MEMBRANE PERMEATION

Norman N. Li

Esso Research and Engineering Company

ABSTRACT

A novel separation technique based on selective permeation
through liquid surfactant membrances for hydrocarbon separations
is described. Some of the possible process schemes are discussed.
The permeation and separation mechanism is discussed in terms of
the effects of temperature, using glycerol and acetone as membrane
additives, solvent-emulsion mixing intensity, and the nature and
composition of feed.

INTRODUCTION

Liquid membrane permeation is a novel separation technique
which gets its selectivity from the relative diffusion rates of
feed components through a liquid surfactant membrane surrounding
a droplet of the feed. It gives nearly perfect membrane selectivity for many hydrocarbons.

This separation technique was discovered from a laboratory
observation that Saponin, a natural surfactant, forms a strong
and visible film at a water/oil interface. The film is so strong
that in a Du Nuoy ring experiment for measuring interfacial tension the ring can actually hook up the film and suspend it in the
top oil phase. The usefulness of this kind of instantly-formed
film for separation purposes was then considered. The first task
was to devise a separation scheme for testing the feasibility of
such an idea. The original separation scheme employed a small
tower, called a liquid membrane diffusion tower. The tower contained three phases--an aqueous surfactant solution at the bottom

section of the tower, a hydrocarbon solvent in the middle section, and a raffinate solution at the top formed by the drops rising out of the solvent phase (this section of the tower was, of course, initially empty). In the tests, drops of a hydrocarbon feed, usually binary mixtures such as n-hexane/benzene, were bubbled through the aqueous surfactant solution from the bottom of the tower. The aqueous liquid membranes formed instantly around the drops and allowed them to rise through the solvent phase without dissolving in it. After emerging from the solvent due to density difference, the drops coalesced in the top section of the column to form the raffinate phase, which was analyzed for enrichment in the non-permeating feed component. The membrane portions of the coalesced drops formed surfactant droplets which descended down to the aqueous bottom phase for recycle.

The initial tests were successful in showing some degree of separation and in proving that the separation scheme was basically workable. Subsequent single drop experiments showed that liquid membranes have nearly perfect selectivity for many hydrocarbons. However, two very serious problems were encountered. One was partial membrane rupture which produced a low overall selectivity in spite of the high inherent selectivity of the membrane. The other was small surface area which resulted in low overall permeation rate. The small surface area was mainly due to large drop size (average drop diameter was about 0.5 cm) and drop sticking in the solvent phase. The sticking of droplets produced large clusters of droplets which reduced the overall surface area for mass transfer. Mixing could not be used to break the clusters and to improve drops-solvent contact because the membranes would be ruptured. Without overcoming these major problems, the liquid membrane separation technique would have remained a laboratory curiosity. The breakthrough came when solutions to these problems were found.

GENERAL DESCRIPTION OF THE MODIFIED "LIQUID MEMBRANE" SEPARATION TECHNIQUE

The first major improvement was to make emulsion-size feed droplets. This involved decreasing the average drop diameter from 0.5 cm to 10^{-3} cm. Both the drop stability and the permeation rate were increased greatly due to the tremendously increased total surface area of the drops. Detailed discussion of this technique has been presented.[5] The other advance was the use of glycerol to strengthen the membrane as discussed in the next section. The combination of these two techniques makes drop breakup insignificant even when the drops are intensively mixed with the solvent. Good separations of many different kinds of hydrocarbon mixtures are therefore achieved. In many cases, separation factors were increased more than 25-fold to values near 100, and permeation rates became much higher than those found in liquid permeation of the same hydrocarbons through 1-mil polymeric membranes.(2,7,8)

Several process schemes are possible with the improved liquid membrane technique. For example, instead of the diffusion columns discussed previously,(5,6) combinations of mixer-and-settler can be used as illustrated in Figure 1. Briefly stated, the feed is emulsified by the surfactant solution recovered from the demulsifier, the emulsion thus made is sent to a mixer to be mixed with the solvent, where selective permeation takes place. The emulsion and the solvent continuously flow into a settler where they are separated into two layers. The solvent phase is sent to a separator to recover the permeates from the solvent; whereas the emulsion is sent to a demulsifier to remove the non-permeable or less-permeable compounds. In addition to the combination of mixer-and-settler, combinations of wetted-wall columns, or diffusion column, emulsifier, and demulsifier may be used.

EXPERIMENTAL APPARATUS AND PROCEDURE

The runs with diffusion columns were described elsewhere. The laboratory scale emulsion runs consisted of first making an emulsion of the feed solution to be separated in an aqueous surfactant solution. The emulsion thus made and the solvent were then fed into another mixer 3" in diameter by 1' long containing a central stirrer which had multiple blades for good mixing. Drop breakup in the mixer was tested by using a dye tracer technique.(5) The mixer content constantly flowed into a settler of same capacity (Figure 2). The mixer-settler unit can be operated either manually or automatically. In the latter case, a liquid level control in each unit and two pumps for feeding respectively the emulsion and the solvent are required. In studying the effects of temperature and the intensity of mixing, batch runs were made with the mixer placed in a heating mantle, which was connected to temperature control instruments.

RESULTS AND DISCUSSION

Elimination of Drop Breakup by Glycerol

Although several surfactant membranes show nearly perfect selectivity for various types of hydrocarbons in single drop experiments, the overall membrane selectivity is low because of drop breakup. Emulsifying the feed droplets does stabilize droplets considerably but does not entirely eliminate breakup. However, glycerol was found to be effective in eliminating drop breakup.

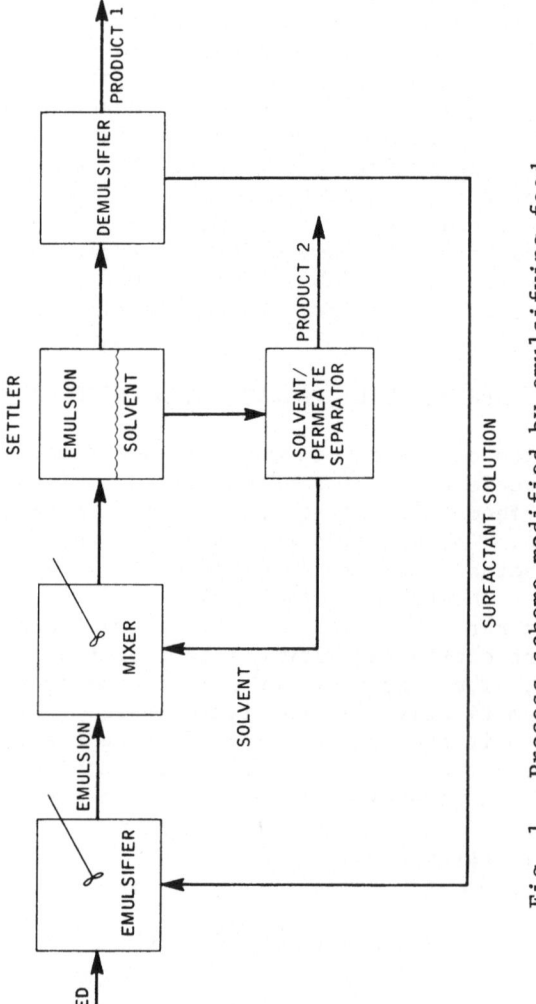

Fig. 1. Process scheme modified by emulsifying feed.

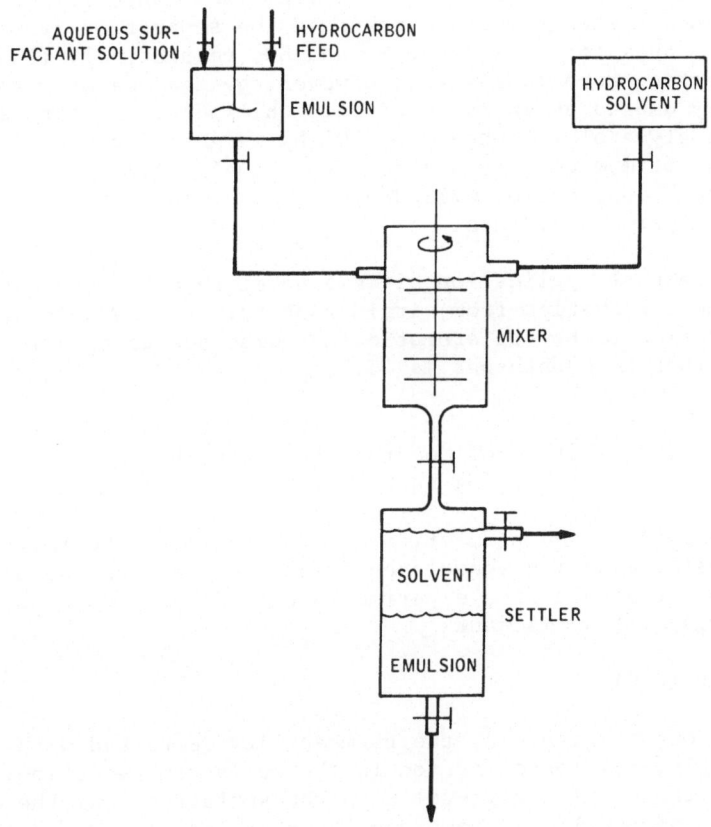

Fig. 2.   Single stage mixer-settler combination.

Adding glycerol to a membrane increases the water layer viscosity and the surfactant layer strength by the synergistic action between surfactants and glycerol. For instance, as shown in Table I, without glycerol, the life of a stationary feed drop (heptane/toluene) coated with a surfactant (Saponin) membrane in a solvent is about 10 min. Adding glycerol extended the drop life to more than three hours. The demulsification method used can be the same in both cases. In actual runs, without glycerol the overall separation factors are 3 for separating benzene from hexane and cyclohexane and 7 for separating benzene from isohexane, isoheptane, and toluene. Using glycerol increased the separation factors in these two cases for benzene to 95 and 60, respectively (Table II). The permeation rate is lowered, however, by the use of glycerol, presumably due to an increase of film thickness. As far as the effect of glycerol is concerned, 70% by weight of glycerol appears to be its optimum concentration. The actual amount of glycerol used in each application will depend on a balance between selectivity and permeation rate.

A potential application of this novel separation technique is to remove aromatics from virgin naphtha. As shown in Table III, the separation factor of aromatics in reference to paraffin is 14.4 and that to naphthenes is 43.7.

Effect of Membrane Thickness and
Thinning Rate on Permeability

Permeability and permeation rate, defined by the following mass transfer equation, have been calculated from the data obtained in the single drop experiments with Saponin used as the membrane-forming surfactant.

$$N = P \ (\overline{C}\text{-}C) \tag{1}$$

As shown in Figure 3, the permeability decreased with increasing glycerol concentration in the surfactant solution. This indicates that mixing glycerol with the surfactant has the effect of thickening the liquid membrane, thereby reducing the diffusion rate. The decrease of permeability was rapid up to the glycerol concentration of 70% at a constant sampling time. This suggests that the membrane had become so viscous at 70% glycerol that further addition of glycerol did not change the membrane thickness appreciably. The film drainage rate became constant at the glycerol concentration of 50% because there was very little change of permeability and permeation rate in 5 minutes.

It should be noted that permeability discussed is different from the one used by some researchers in describing diffusing through a monolayer.[1,4] A mathematical analysis is made to il-

TABLE I

EFFECT OF GLYCEROL ON MEMBRANE LIFE AND PERMEATION RATE

Surfactant: Saponin
Solvent for Permeates: Hexane
Feed: Toluene/Heptane (A_7/nC_7)
                            (1/1 by wt.)
Temperature: 25°C

Solvent for Saponin	Water	Water + Glycerol (30%)	Water + Glycerol (70%)
Membrane Life (Life of Feed Drop) (Min.)	10	194	200
Permeation Rate of A_7 (gm/hr., 100 cm^2)	12.1	4.2	0.80
Time for nC_7 to Appear in Solvent (Min.)	10	50	85

lustrate such difference for the simple case of a solute diffusing from a homogeneous phase, through a monolayer, into another phase.

As shown in Figure 4, the diffusion is driven by an overall concentration difference $\triangle C$, which is C_1-C_2.

In a steady state, if the solute flux across the total distance Z_t is described by the Fick's Law, it is

$$\left(\frac{dN}{dt}\right)_t = DA\ \frac{C_1-C_2}{Z_t} \tag{2}$$

Since Henry's Law for the monolayer can be written as:

$$C_1' = H_1C_1 \text{ and } C_2' = H_2\overline{C}_2 \tag{3}$$

The overall diffusion can be broken down into separate diffusion steps for the surfactant monolayer and the solvent phase No. 2.

TABLE II

EFFECT OF GLYCEROL CONCENTRATION ON SEPARATION FACTOR

Run Temperature = 25°C

| | Permeate | Separation Factors of Permeate (Referred to Paraffin) at Different Glycerol Concentrations in Water | | |
		0%	30%	70%
(1) Surfactant: Saponin (0.2%)				
Feed I	A_6	3.7	5.8	14.3
	A_7	2.9	5.8	8.5
Feed II	A_6	6.7	8.4	60.0
	A_7	4.7	5.0	32.1
Feed III	A_6	2.9	3.6	95.0
	Cyclo C_6	1.2	1.2	2.7
Feed IV	$C_6=$	1.5	-	2.0
	$C_6==$	2.0	-	4.0
(2) Surfactant: Dodecyl Sodium Sulfate (0.2%)				
Feed I	A_6	4.4	5.3	13.4
	A_7	2.6	2.5	7.7
Feed II	A_6	8.4	7.0	28.8
	A_7	5.4	3.9	15.0

Feed I: Hexane/Benzene/Toluene

Feed II: 2,4-Dimethyl Butane/2,4-Dimethyl Pentane/Benzene/Toluene

Feed III: $nC_6/A_6/cyclo\ C_6$

Feed IV: $nC_6/1-C_6^=/1,5\ C_6^{==}$

TABLE III

SEPARATION OF AROMATICS FROM VIRGIN NAPHTHA

Surfactant Solution: 0.1% Saponin + 70% Glycerol + Water

Run Temperature: 25°C

Components	Feed Composition (Wt. %)	Permeate Composition in Solvent Phase (Wt. %)
Total Aromatics	21.2	85.3
Total Paraffins	39.8	11.1
Total Naphthenes	39.0	3.6

Separation Factors of Aromatics:

   (1)   Based on Paraffins = 14.4.

   (2)   Based on Naphthenes = 43.7.

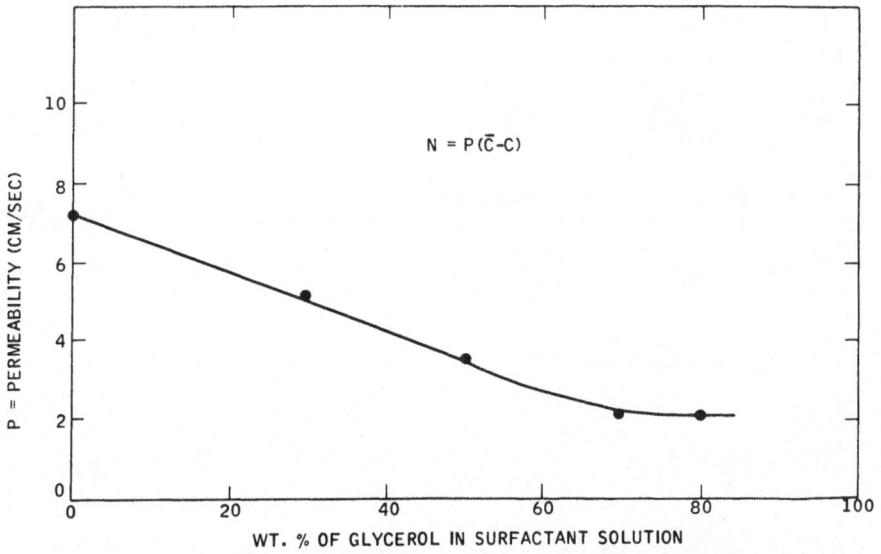

Fig. 3.   Effect of glycerol concentration on permeability.

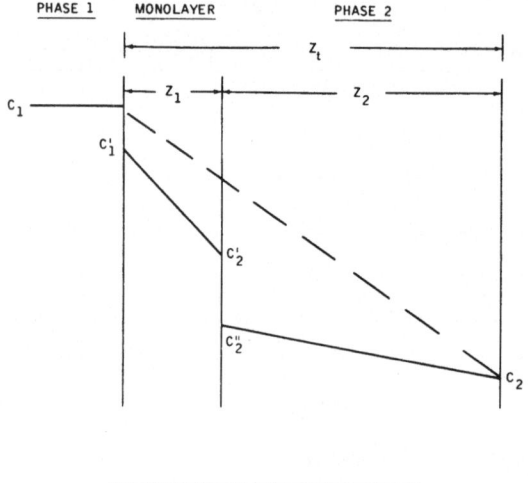

Fig. 4. Concentration gradients for diffusion through a monolayer.

$$\left(\frac{dN}{dt}\right)_1 = D_1 A \frac{C_1'-C_2'}{Z_1} \tag{4}$$

$$\left(\frac{dN}{dt}\right)_2 = D_2 A \frac{C_2''-C_2}{Z_2} \tag{5}$$

In steady state, the flux as given by Equations (2), (4), and (5) are equal, i.e.,

$$\left(\frac{dN}{dt}\right)_t = \left(\frac{dN}{dt}\right)_1 = \left(\frac{dN}{dt}\right)_2 \tag{6}$$

From the above equations, we obtain

$$\left(\frac{D_1}{A_1}\right) H = \frac{(D_2/A_2)(D_t/A_c)}{(D_2/A_2) - (D_t/A_t)} \tag{7}$$

$$\text{or } P_1' H = \frac{P_2' P_t'}{P_2 - P_t} \tag{8}$$

$$\text{where } P_1' = D_1/A_1 \tag{9}$$

$$P_2' = D_2/A_2 \tag{10}$$

$$P_t' = D_t/A_t \tag{11}$$

P_1', P_2', and P_t' are, therefore, Fick's Law permeabilities.

To relate these parameters to the liquid membrane permeability, or overall mass transfer coefficient, one can write the following equation by combining Equations (1) and (2)

$$\left(\frac{dN}{dt}\right)_t = P_t (C_1 - C_2) = P(\overline{C}_1 - C_1) \qquad (12)$$

From Equations (8) and (12), one obtains

$$P = \frac{P_1' P_2' H}{P_1' H + P_2'} \left(\frac{\overline{C}_1 - C_2}{\overline{C}_1 - C_1}\right) \qquad (13)$$

Relationship Between Film Drainage and Weight Ratio of Surfactant Solution to Feed

The minimum weight ratio of surfactant solution/feed necessary for forming a stable emulsion can be determined by two methods. One is to make a series of emulsification tests at varying surfactant solution to feed ratios. The other is a drainage experiment, which involves making an emulsion with a much higher than necessary surfactant solution-to-feed ratio and measuring the volume of the surfactant solution drained from the emulsion phase under the influence of gravity to determine the final amount of surfactant solution retained in the emulsion phase. Close agreement between these two methods can usually be obtained. For example, for the typical case where the feed was a mixture of 1/1 toluene/heptane and the surfactant solution was composed of 0.2% Saponin, 29.8% water and 70% glycerol. The minimum weight ratio was determined to be 0.45 from the emulsification tests and to be 0.45 from the drainage experiment starting from a ratio of 1.2 (Figure 5). In general, the second method, being much easier to use than the first, is employed to determine the minimum amount of surfactant solution needed to form a stable emulsion for a given feed-surfactant system. It should be noted that the drainage curves obtained when plotted as the amount of surfactant ·solution drained vs. time can be closely approximated by the commonly used drainage equations such as the equation recommended by Davies and Rideal for slowly draining foams.[3]

$$\frac{dv}{dt} = \frac{BVo}{2} (Bt + 1)^{-3/2} \qquad (14)$$

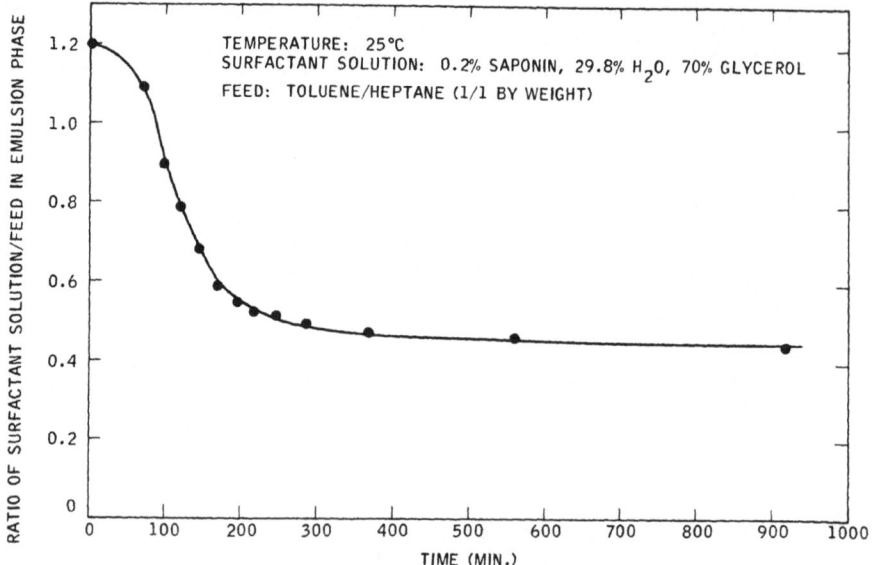

Fig. 5.   Drainage of surfactant solution from a liquid membrane
emulsion.

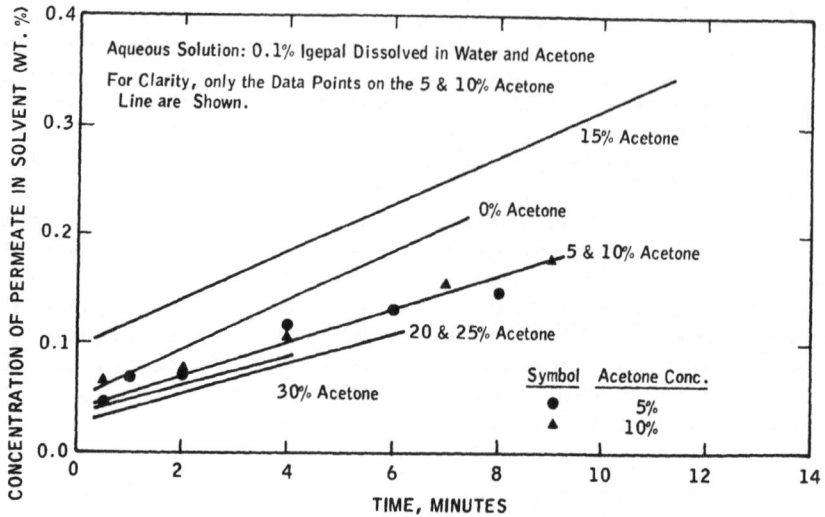

Fig. 6.   Permeation of toluene through igepal film.

Effect of Acetone in Liquid Membrane
on Membrane Selectivity

To examine further how the permeation rate is influenced by additives in the water layer in a membrane, mixed solvents of water and acetone were used to dissolve surfactants. Acetone was chosen because of its good miscibility with water, as well as with the hydrocarbon feed. It was hoped, therefore, that the use of acetone in the membrane might increase the permeation rate of hydrocarbons. However, the diffusion rates of toluene and heptane through the liquid membranes formed in such a solution were found to remain constant throughout the acetone concentration range from 0 to 30%, indicating that the presence of acetone in the water layer did not facilitate the transfer of hydrocarbons. This is shown by the relatively unchanged slopes of permeate concentration vs. time curves in Figure 6. The difference between the starting points of any two curves was chiefly due to feed contamination on the walls of the container used. At acetone concentrations higher than 30%, the membranes around the toluene/heptane drops became highly unstable and ruptured practically at the moment they were formed.

Although the exact reasons for the decrease of membrane stability when acetone is added to the membrane are not yet completely clear, viscosity appears to be a major influencing factor. Membrane viscosity and thickness are interrelated because the more viscous the surfactant solution, the thicker the membrane. Thicker membranes give rise to lower permeation rate but higher drop stabilities. The following table shows that adding acetone to water will lower the water layer viscosity and, hence, increase the film drainage rate, whereas adding glycerol to water increases the water layer viscosity and thus increases the film stability.

Temperature: 60°C

Compound	Viscosity (Centipoise)
Acetone	0.24
Water	0.50
Glycerol	100.0

Membrane Structure and Temperature Effect

The permeation rate was found to increase exponentially with temperature (Figure 7), whereas the selectivity remained about the same throughout the temperature range studied. As discussed previously,[5,6] the liquid membrane on the surface of a hydrocarbon drop in a hydrocarbon solvent should have, by necessity,

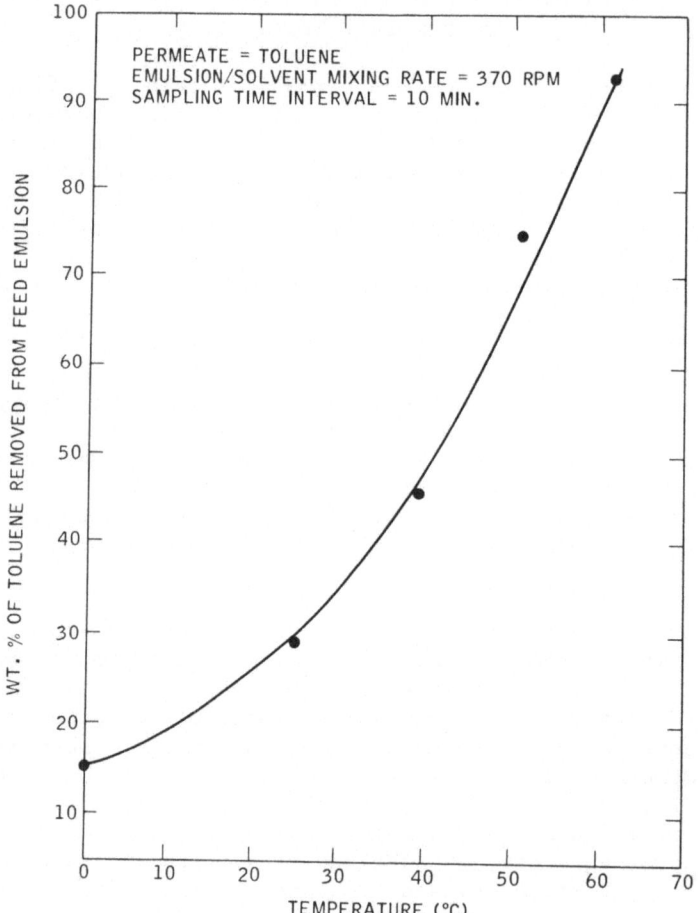

Fig. 7. Effect of temperature on permeation rate.

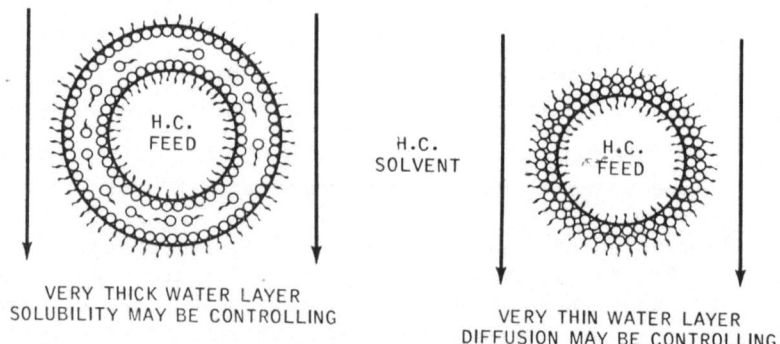

Fig. 8. Structure of a liquid membrane (H.C.= hydrocarbon).

two surfactant layers with a water layer containing some free
surfactant molecules in between (Figure 8). When the drop undergoes violent mixing motion in the solvent phase, the outer layer
of surfactant and portion of the inner water layer are constantly
being wiped off the membrane. New outer layer of surfactant
apparently can be formed almost instantly by the free surfactant
molecules in the water layer. The water layer may eventually
become ultra-thin and tightly bonded to the hydrophilic ends of
the two surfactant layers. For this kind of membrane structure
of two surfactant layers with one water layer in between, the
permeation rate of a permeate is a function of its solubility in
and diffusivity through the membrane. The diffusivities of hydrocarbons through a liquid membrane should be quite close in value
as suggested from the known diffusivities of hydrocarbons through
water. In comparison, the solubilities of hydrocarbons in water
are quite different.[9] The solubility of a hydrocarbon in water
is therefore the controlling factor in determining its permeation
rate and the membrane selectivity.[5,6] In the case of varying
temperature in studying permeation rate, the diffusivity usually
increases approximately linearly with temperature, and the solubility varies to a much lesser degree with temperature. The decrease of membrane thickness, a third factor, therefore appears
to be the main factor which accounts for the rapid increase of
permeation rate with temperature. This is shown by the rate vs.
time data at constant temperature in single drop studies, where
the rate always sharply increases near the time of membrane rupture due to membrane thinning effects. For the hydrocarbons
studied, their ratios of solubilities in water change very little
with temperature within the temperature range studied. This accounts for the relatively constancy of membrane selectivity
throughout the temperature range studied. It should be noted that
membrane selectivity does not correspond to volatility difference.
The two examples given in Table IV show the more permeable components are less volatile (higher boiling point) than the less
permeable components.

Because of the reason that a membrane becomes thinner at
higher temperature due to higher drainage rate, the highest temperature one can study is limited by the membrane stability.
This means that at a certain high temperature, the membrane becomes so unstable that complete membrane rupture occurs. In an
actual process, a balancing point needs to be determined by
economics studies, at which the cost increase due to temperature
increase is offset by the gain in rate increase.

Effect of Mixing Intensity

Since the function of the solvent used in contacting the
emulsion is to remove the permeates, intimate contact of solvent

TABLE IV

MEMBRANE SEPARATION DOES NOT DEPEND ON VOLATILITY DIFFERENCE

Hydrocarbon Mixture	B.P. (°C)	Preferentially Diffusing Comp.	Separation Per Stage
I. nC_7	98		
A_7	110	A_7	50% — 99%
II. nC_6	68		
Cyclo C_6	81	Cyclo C_6	50% — 76%

TABLE V

EFFECT OF FEED COMPOSITION ON SEPARATION FACTOR

Run Temperature: 25°C

Feed Composition (A_7/nC_7)	Permeate Composition In Solvent (A_7/nC_7)	Separation Factor of A_7 In Reference to nC_7
20/80	55/45	22
50/50	79/21	51
70/30	96/4	91

with feed drops is therefore important. However, because of the
surface compatibility among the drops and the density difference
between the drops and the solvent, the drops tend to stick together. This reduces the total surface area for mass transfer.
Good mixing of the emulsion with the solvent is therefore essential in obtaining high permeation rate. This cannot be done if
liquid membranes were weak as intensive mixing would break up the
membranes. However, since the membrane stability can be increased
tremendously by making emulsion-size droplets and by the use of
glycerol as a membrane strengthening agent, intensive mixing can
be applied. As shown in Figure 9 and Figure 10, both the rate
and the separation factor obtained in batch runs increased with
increasing stirring speed. The fact that separation factor was
also increased shows that drop break up was insignificant, if not
zero. It was mainly the consequence of the increase of permeation
rates of permeates since separation in this membrane technique is
entirely based on rate difference. For the same reason, both the
separation factor and permeation rate obtained in batch runs will
decrease at time longer than shown in the drawings. They will
reach a constant value instead of declining in continuous runs.

Effect of a Feed Composition
on Membrane Selectivity

Both membrane selectivity and permeate composition vary with
feed composition. The results in Table V and the previously published results[5,6] show:

(1) The above conclusion is partly due to the fact that membrane structure and thickness vary as a function of the nature of
feed because of structure compatibility.

(2) Solubility is a more important controlling factor than
diffusivity for permeation as discussed in detail elsewhere.[5]
However, the effect of diffusivity is still appreciable. If
solubility alone were controlling, the selectivity would be independent of the feed composition. If diffusivity alone were controlling, selectivity would vary with the feed composition in such
a way as to maintain a constant permeate composition.[7]

Although membrane selectivity is largely dependent on solubility, it should be noted that it is not the absolute values of
the solubilities of the permeates, but the relative solubilities,
or solubility ratio that determines the relative selectivity. The
uniqueness of this novel separation technique, therefore, lies in
utilizing water (or other solvent for surfactants) as an ultra-
thin membrane. If water were used as the solvent in extraction
for hydrocarbon separations, a huge amount of water will be needed
to get appreciable yield and a tedious separation of the hydrocar-

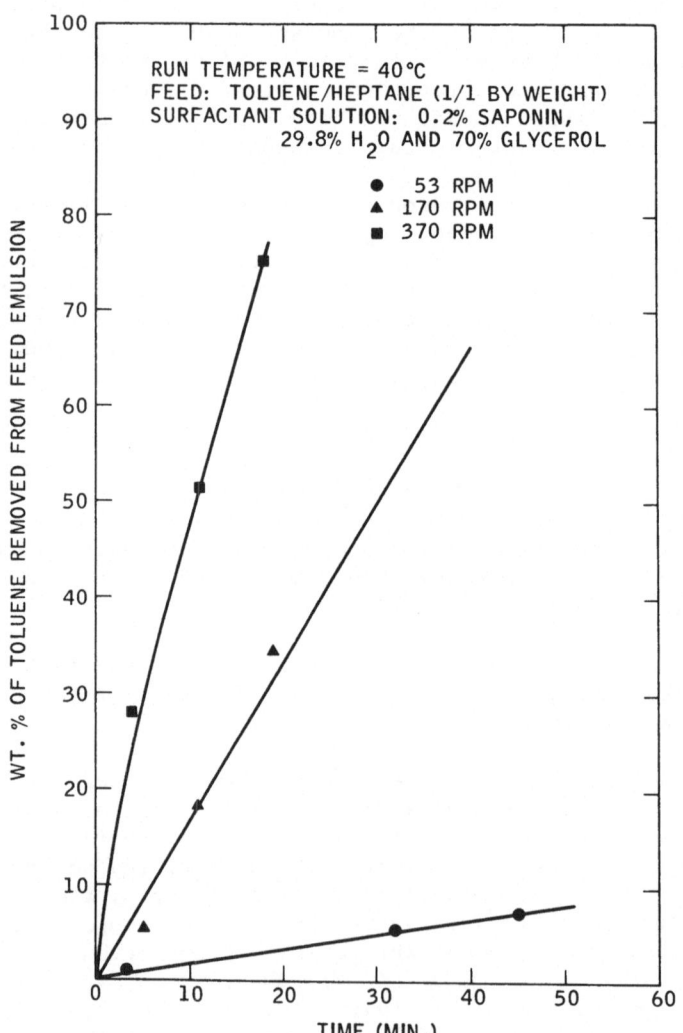

Fig. 9. Effect of mixing intensity on permeation rate.

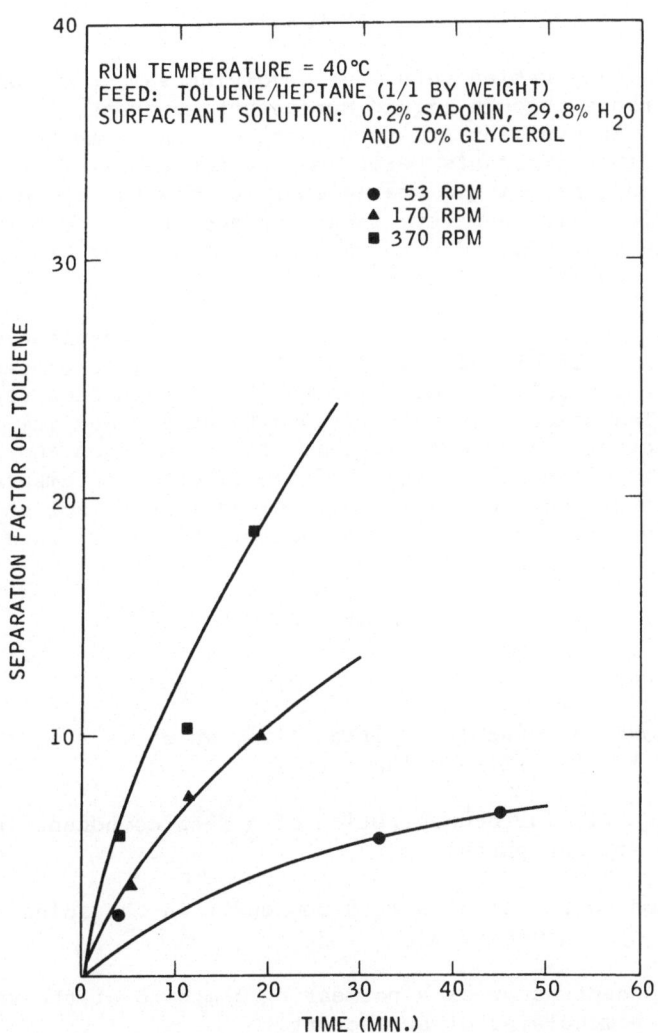

Fig. 10. Effect of mixing intensity on separation factor.

bons, as solutes, from water is also necessary; whereas in a liquid membrane process, very little water is needed to achieve the same yield and there is no problem of separating permeates from water because the permeates automatically diffuse out of the water in the membranes due to the imposed concentration gradient.

SUMMARY

A novel separation technique involving selective permeation of hydrocarbons through liquid surfactant membranes has been developed. An emulsion-treating technique and a method of using glycerol with surfactants before making the emulsion are very effective in generating a stable emulsion having liquid membranes of high selectivity and very large surface area. Such an emulsion can be intensively mixed with wash solvent without drop break up and without emulsifying the solvent.

The historical background as to how this technique was invented is briefly discussed. Some of the possible process schemes using diffusion columns and mixer-settler combination are described. The unique features of this technique and its comparison with extraction are briefly stated. The permeation and separation mechanism is discussed in terms of the effects of temperature, using glycerol and acetone as membrane additives, solvent-emulsion mixing intensity, and the nature and composition of feed.

NOTATION

A = Interfacial area.

B = Constant used in Equation (14), which is a measure of initial film drainage rate.

$\bar{C}$ = Equilibrium concentration of a feed component in the solvent phase.

C = Concentration of a feed component in the solvent phase at a given time.

C' = Concentration of a permeating compound at the surface of a monolayer of surfactant.

C'' = Concentration of a permeating compound in the solvent phase near a monolayer of surfactant.

D = Diffusion coefficient.

H = Henry's law constant.

N = Permeation rate of a feed component (gm./sec., cm^2).

P = Membrane permeability (cm./sec.) corresponding to an overall mass transfer coefficient.

P' = Permeability used in Fick's law diffusion equations.

t = Time.

V = Volume of the surfactant solution drained from a column of emulsion.

V_o = Original volume of a surfactant solution in a column of emulsion.

Z = Distance of permeation.

Subscripts

1 = In the surfactant monolayer.

2 = In the solvent phase.

t = For the total system including a surfactant monolayer and solvent phases.

LITERATURE CITED

1. Adamson, A. W., "Physical Chemistry of Surfaces," New York, Interscience Publishers, Inc., 1967.

2. Binning, R. C., R. J. Lee, J. F. Jennings, E. C. Martin, Ind. Eng. Chem., 53, 45 (1961).

3. Davies, J. T., and E. K. Rideal, "Interfacial Phenomena," Academic Press, New York, 1963.

4. LaMer, Y. K., ed., "Retardation of Evaporation by Monolayers," Academic Press, New York (1962).

5. Li, N. N., A. I. Ch. E. J, 17, (1), (1971) (In Press).

6. Li, N. N., Preprint Volume of Material Engineering and Science Division Biennial Conference p. 291, 67, A.I.Ch.E. National Meeting, Atlanta, Georgia, February 15-18, 1970.

7. Li, N. N. and R. B. Long, "Membrane Processes" Perry's Chemical Engineering Handbook, 5th Edition, McGraw Hill, New York (In Press).

8. Li, N. N. and R. B. Long, In "Progress in Separation and Puri-
    fication" Vol. 3 (E. S. Perry, ed.), Interscience, New York
    (In Press).

9. McAuliffe, C., J. Phys. Chem., $\underline{70}$, 1267 (1966).

ULTRAFILTRATION

Alan S. Michaels, Lita Nelsen, Mark C. Porter

Amicon Corporation, Lexington, Massachusetts

I.   THE PROCESS OF ULTRAFILTRATION

"Ultrafiltration" is a process of separation whereby a solution
containing a solute of molecular dimensions significantly greater
than that of the solvent is depleted of solute by being forced
under an hydraulic pressure gradient to flow through a suitable
membrane. "Reverse osmosis", ultrafiltration, and ordinary filtration differ superficially only in the size scale of the particles which are separated; differentiation between the three is in
large measure arbitrary. It is, however, convenient to reserve
the term "reverse osmosis" for membrane separations involving solutes whose molecular dimensions are within one order of magnitude
of those of the solvent, and to use "ultrafiltration" to describe
separations involving solutes of molecular dimensions greater than
ten solvent molecular diameters, and below the limit of resolution
of the optical microscope (ca. 0.5 μ). "Ultrafiltration" thus
encompasses all membrane moderated, pressure activated separations
involving solutions of modest molecular weight (ca. 500 and up)
solutes, macromolecules, and colloids. At present, ultrafiltration processes are largely confined to aqueous media, and most of
what follows relates to aqueous systems. There are, however, no
fundamental reasons why ultrafiltration cannot be performed with
non-aqueous solvents (utilizing, of course, solvent resistant
membranes); as a matter of fact, there are numerous commercially
important petroleum and petrochemical purifications which can and
ultimately will be performed by ultrafiltration with suitably con-
stituted membranes.

Ultrafiltration can be utilized to accomplish one or more of the
following: (1) concentration of solute by removal of solvent;

(2) <u>purification</u> of <u>solvent</u> by removal of solute; (3) separation
of Solute A from Solute B by ultrafiltration through a membrane
permeable to A but not to B (or vice versa); and (4) analysis of
a complex solution for specific solutes to which the membrane is
permeable.

While each of these procedures can be performed by alternate means
(e.g., evaporation, dialysis, ultracentrifugation, chemical precipitation, etc.), ultrafiltration is usually the method of choice
from the point of view of speed, efficiency, and cost, particularly
when thermally unstable or biologically active materials are involved, or when large volumes of dilute solutions are to be processed.

The basic ultrafiltration process is extremely simple, as illustrated in Figure 1. Solution to be ultrafiltered is confined
under pressure (utilizing either compressed gas or a liquid pump)
in a cell, in contact with an appropriate ultrafiltration membrane
supported on a porous plate. The contents of the cell must be
subjected to moderate agitation to avoid accumulation of retained
solute on the membrane surface with attendant "blinding" of the
membrane. Ultrafiltrate is continuously produced and collected
until the retained solute concentration in the cell solution
reaches the desired value. With a high quality ultrafiltration
membrane and adequate agitation, the solute concentration in the
"retentate" can often be allowed to rise as high as 20-50% by
weight of the solution before the ultrafiltration rate begins to
decrease significantly. Since many ultrafiltrations are carried
out on quite dilute (e.g., 0.01-1% solids by weight) solutions,
concentration factors as large as 1,000, or solvent recoveries
as high as 99%, can be accomplished without difficulty. For the
same reason, virtually complete separation of retained solutes
from unretained solutes can often be accomplished by direct ultrafiltration.

When the primary objective of ultrafiltration is to separate virtually quantitatively a membrane permeable solute (such as salt)
from a membrane impermeable solute (such as a protein), the process of "pressure dialysis" or "diafiltration" is usually employed.
In this process, the solution to be purified is confined under
pressure in contact with the ultrafiltration membrane, and pure
solvent is added to solution at the same rate as ultrafiltrate is
removed through the membrane. Under these circumstances, the concentration of membrane impermeable solute remains constant, while
the concentration of permeable solutes decreases exponentially
with the volume of solvent delivered to the system. In this
fashion, salts and other microsolutes can be essentially completely
eliminated from solution rapidly, with minimal consumption of dialyzing solvent, and without product dilution.

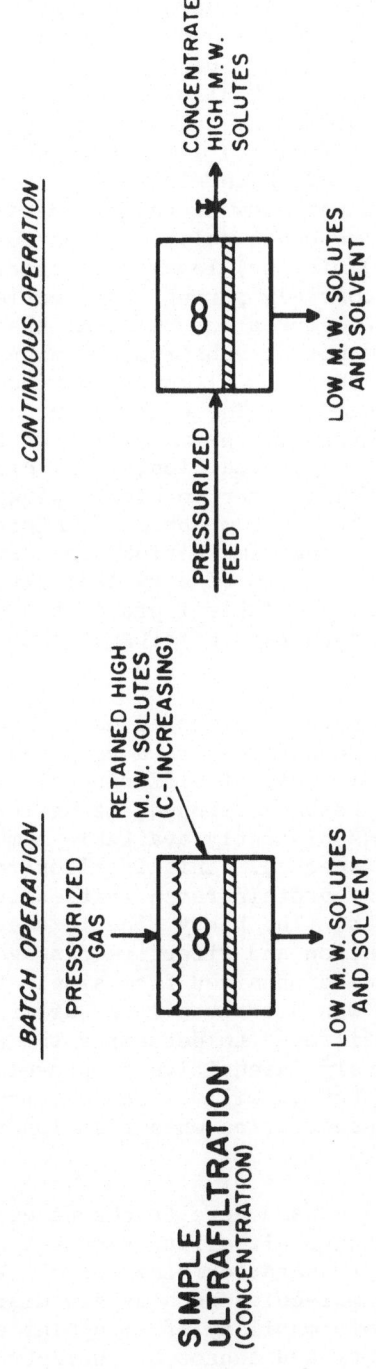

Fig. 1. Processes employing ultrafiltration membranes.

II.   ULTRAFILTRATION MEMBRANES

The emergence of ultrafiltration as a practical molecular separation process for both laboratory and industrial use has been a
consequence of the development, over the past five years, of an
expanding family of so-called "anisotropic" or "asymmetric" semipermeable polymeric membranes covering a broad range of pore sizes
of molecular dimensions (from ca. 10 to ca. 200 Angstroms) and narrow pore size distribution, which display extraordinarily high
water permeabilities at very low operating pressures (10-100 psi).
These membranes are microporous polymer films comprising an exceedingly thin (0.1-2.0 microns) layer of ultrafine pore material
overlying a much more coarsely porous, open celled foam; the very
thin "skin" on these structures functions as the molecular separation barrier, while the porous substructure provides the necessary
mechanical support for the skin without contributing significantly
to the fluid flow resistance. The extreme thinness of the skin
accounts for the high hydraulic permeability of the membrane despite its small pore size; in addition, it is difficult for the
pores in the very thin skin layer to become plugged or blocked by
solute molecules, so that these membranes exhibit sustained high
liquid throughputs over long time periods in service. An electron
micrograph of the cross section of a typical ultrafiltration membrane is shown in Figure 2. Table I presents a summary of the
principal properties and characteristics of commercially available
membranes.

An ideal ultrafiltration membrane would have an exceedingly narrow
pore size distribution, and therefore a very sharp "cutoff" in its
capacity to permeate molecules of differing size. While such an
ideal membrane remains to be developed, it is surprising, and fortunate, that many of the presently available membranes display remarkably narrow cutoff spectra. This is illustrated in Table II
and Figure 3, which show protein retentivities for two classes of
Amicon Diaflo® membranes. The heavy line in the table delineates
between effective retention and effective passage of protein by a
membrane. These data were obtained from single solute studies
where a single protein was dissolved in a buffer solution, and
the solution was ultrafiltered to determine the retentivity. Retentivities are relatively insensitive to protein concentration in
single solute studies, but as will be shown below, they are affected adversely by concentration when fractionating a mixture of
proteins.

Molecular weight has been used here to characterize the size of
molecules as they influence ultrafiltration retentivity. This is
a satisfactory method of characterizing the ultrafilterability of
a homologous series of molecules such as globular proteins, but
the size, shape, and deformability of molecules each influence ultrafiltration retentivity and cannot be incorporated into a single

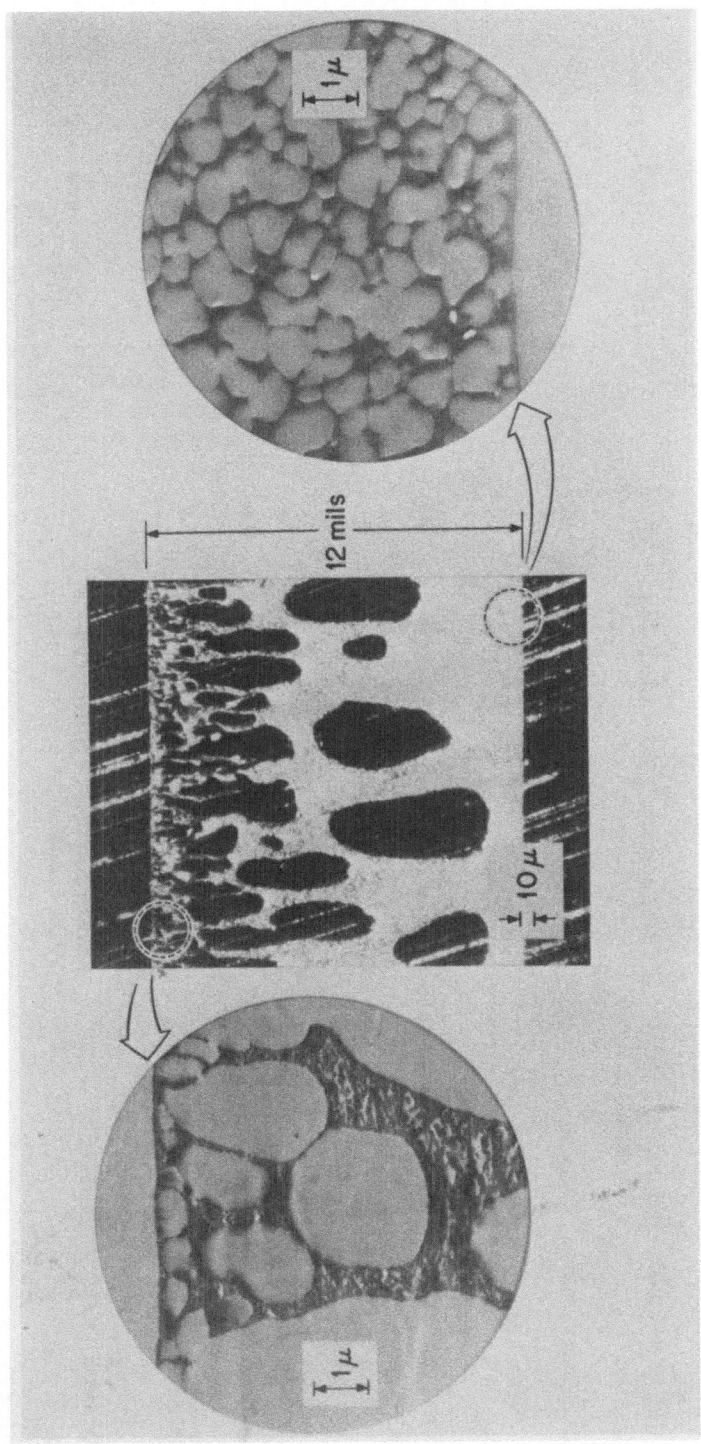

Fig. 2. Cross section of a typical ultrafiltration membrane.

TABLE I
COMMERCIALLY AVAILABLE ULTRAFILTRATION MEMBRANES

Description	Manufacturer	Type	Water Permeability at 100 psi (ml/cm^2-min)	Molecular Weight Cutoff (80-100% Retention)
Gel Cellophane	duPont Union Carbide	Homogeneous (?) Cellulosic	0.004	10,000 (Cytochrome C)
PEM Membrane	Gelman	Isotropic Cellulosic	0.02	40,000 (Ovalbumin)
"P-Membrane"	Schleicher & Schuell	Homogeneous Cellulosic	0.08	60,000 (Albumin)
CA-Type B	General Atomics	Anisotropic, Cellulose Acetate	0.007	600 (Raffinose)
CA-Type C	General Atomics	Anisotropic, Cellulose Acetate	0.003	350 (Sucrose)
Diaflo UM-05*	Amicon	Anisotropic, Polyelectrolyte Complex	0.05	350 (Sucrose)
Diaflo UM-2	Amicon	Anisotropic, Polyelectrolyte Complex	0.1	600 (Raffinose)
Diaflo UM-10	Amicon	Anisotropic, Polyelectrolyte Complex	0.3	10,000 (Dextran 10)
Diaflo PM-10	Amicon	Anisotropic, Aromatic Polymer	0.5	10,000 (Cytochrome C)

* Ion exchange membrane

TABLE I, continued

Description	Manufacturer	Type	Water Permeability at 100 psi (ml/cm^2-min)	Molecular Weight Cutoff (80-100% Retention)
Diaflo PM-30	Amicon	Anisotropic, Aromatic Polymer	0.7	30,000 (Ovalbumin)
Diaflo XM-50	Amicon	Anisotropic, Substituted Olefin	0.7	50,000 (Albumin)
Diaflo XM-100A	Amicon	Anisotropic, Substituted Olefin	0.9	100,000 (7S Globulin)
Diaflo XM-300	Amicon	Anisotropic, Substituted Olefin	1.1	300,000 (Apoferritin)
HFA-100	Abcor	Anisotropic, Cellulosic	0.07	10,000 (Dextran 10)
HFA-200	Abcor	Anisotropic, Cellulosic	0.40	20,000 (Dextran 20)
HFA-300	Abcor	Anisotropic, Cellulosic	1.40	70,000 (Albumin)
PSAC	Millipore	Anisotropic, Cellulosic	1.2	1,000 (Grom Cresol Green)
PSED	Millipore	Anisotropic, Cellulosic	0.75	25,000 (α Chymotrypsin)
PSDM	Millipore	Anisotropic, Cellulosic	1.0	40,000 (Ovalbumin)

TABLE II

PROTEIN RETENTIVITIES OF DIAFLO® ULTRAFILTRATION MEMBRANES

Compound	Molecular Weight	% Retention				
		XM-300 $(300,000)^1$	XM-50 (60,000)	XM-4A (35,000)	UM-10 (10,000)	UM-2 (1,000)
Apo-Ferritin	480,000	100^2	100	---	---	---
Gamma Globulins	160,000	100^2	100	---	---	---
Aldolase	142,000	0	85	100	100	100
Albumin	67,000	0	68	100	100	100
Hemoglobin	64,500	0	62^2	70	100	100
Ovalbumin	45,000	0	77	98	100	100
Chymotrypsinogen A	25,000	0	0	0	90	100
Alpha-Chymotrypsin	24,500	0	0	0	95	100
Trypsin	20,000	0	0	0	95	100
Myoglobin	17,800	0	0	0	88	100
Cytochrome C	12,400	0	0	0	85	100
Cyanocobalamin	1,355	0	0	0	0	0
Phenylalanine	165	0	0	0	0	0

[1] Values in parentheses represent arbitrary cutoff limits ascribed to the various membranes.

[2] Membrane plugging was observed with these compounds as evidenced by a considerable increase in pressure during ultrafiltration. This was not observed in the tighter membranes. Membrane plugging of this membrane can be eliminated by reducing the ultrafiltration rate, and thus the pressure.

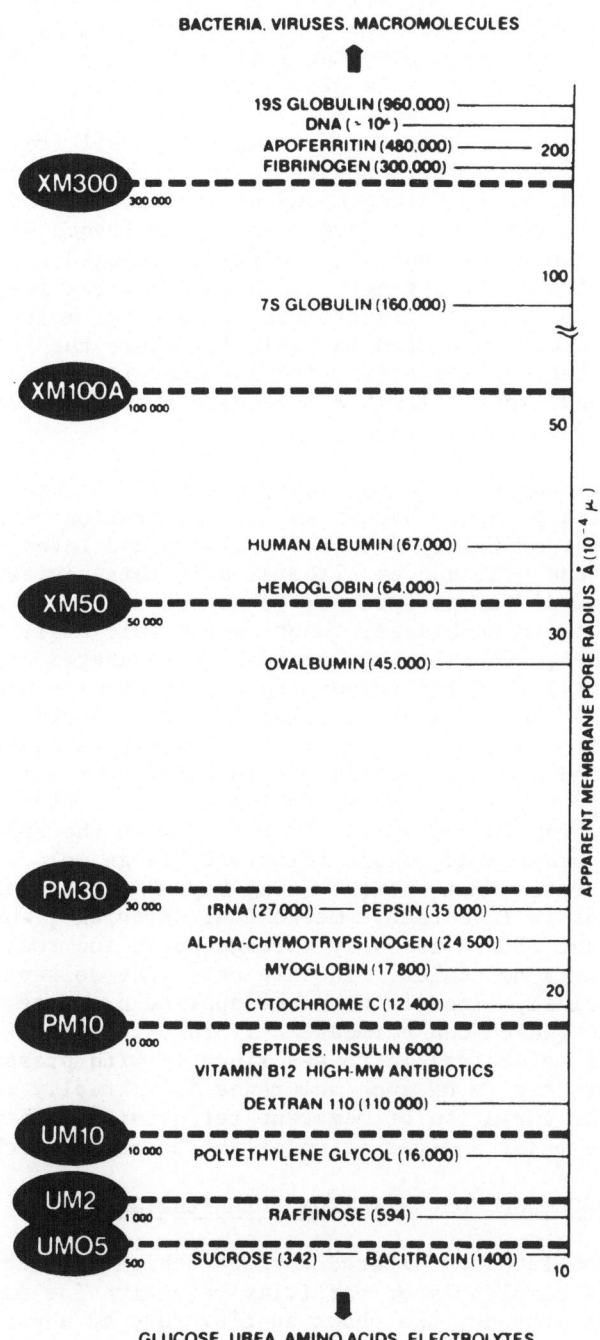

Fig. 3. Protein retentivities for AMICON diaflo membranes.

parameter such as molecular weight. As a general rule linear, flexible molecules in solutions ("free draining chains") tend to be retained to a lesser degree for a given molecular weight than do more highly structured molecules such as branched polymers of globular proteins. Apparently the fluid shear stresses in the vicinity of membrane pores are high enough to uncoil free draining chains and to reduce their effective ultrafiltration cross sectional area. The ionic strength and pH of solutions of polyelectrolytes being ultrafiltered have a strong influence on their retentivities. The more a polyelectrolyte is charged in solution, and the lower the ionic strength of the medium, the larger is the effective size of the polyelectrolyte for a given molecular weight. These factors are illustrated in Table III where the relative ultrafiltration retentivities for globular proteins, branched polysaccharides, and linear flexible molecules of various molecular weight are shown.

The molecular transport mechanisms by which solute and solvent are separated during passage through an ultrafiltration membrane have been the subject of fairly detailed analysis and interpretation elsewhere (1) and will not be elaborated in this review. Suffice to say that two different types of transport mechanism are believed to operate in these membranes. Where relatively small (molecular weight ca. 500-2,000) solute molecules are separated using "tight" membranes, both solute and solvent appear to migrate through the membrane by the processes of dissolution and molecular diffusion; under these circumstances, the solvent flux varies nearly linearly with the applied pressure, while the solute flux remains nearly constant. Such membranes display solute retentivities or rejection coefficients which increase hyperbolically with the applied pressure. On the other hand, where relatively large molecules or colloids are separated from solvent using "open" membranes, the separation process is in effect a mechanical "sieving", whereby solute free solvent flows viscously through pores too small to accommodate the solute molecules. In this case, the solvent flux increases essentially linearly with the applied pressure, while the flux of solute (which can permeate only through pores large enough to pass solute molecules) increases linearly with pressure also; hence, the retentivity of such membranes is virtually pressure independent. The variation of membrane retentivity with pressure is, therefore, the major manifestation of the transport mechanism.

III. THE PROBLEM OF CONCENTRATION POLARIZATION

When a solution is ultrafiltered through a high permeability membrane which is completely or partially retentive for one or more of the solutes present, the observed flux rate at any applied pressure is invariably lower, and often far lower, than the value measured with pure solvent at the corresponding pressure. This flux reduction is not attributable to "plugging" of the ultrafiltration

TABLE III

EFFECT OF SIZE AND SHAPE OF MOLECULES ON ULTRAFILTRATION RETENTIVITY

Membrane*	Solute Class		
	Globular Proteins	Branched Polysaccharides	Linear, Flexible Polymers
	γ-Globulin (160,000)**		
Diaflo XM-50	Albumin (69,000)		
Diaflo PM-30	Pepsin (35,000)	Dextran 250 (236,000)	
	Cytochrome C (13,000)		Polyacrylic Acid (pH 10; 50,000)
	Insulin (5,700)		
Diaflo PM-10		Dextran 110 (100,000)	Polyacrylic Acid (pH 7; 50,000)
	Bacitracin (1,400)	Dextran 40 (40,000)	
Diaflo UM-10		Dextran 10 (10,000)	Polyethylene Glycol (20,000)

* Molecules above a horizontal membrane line are completely retained by the membrane; below the line partial retention or complete clearance is observed.

** Numbers in parentheses denote molecular weights.

membrane, but rather to the accumulation, on the upstream surface of the membrane, of a layer of highly concentrated solution of retained solute convectively transported to the membrane by the flow of solvent through the membrane. This is the phenomenon of "concentration polarization", which is well recognized as a problem in water demineralization by reverse osmosis, and has been a subject of extensive study in that field. However, the flux diminution arising from concentration polarization with macrosolute solutions during ultrafiltration is not due to osmotic pressure effects as in reverse osmosis, since the osmotic pressures of most macrosolutes are negligibly small at all practicable concentration levels. Thus, another explanation must be sought for the observed flux reduction in ultrafiltration.

The physics of the concentration polarization process are illustrated in Figure 4. As the solution is ultrafiltered, solute is brought to the membrane surface by convective transport according to the equation:

$$J_{p_{to}} = J_w \bar{c} \tag{1}$$

where: $J_{p_{to}}$ is the rate of arrival of retained solute at the membrane surface (gm/cm^2-min); J_w is the ultrafiltrate rate (cc/cm^2-min); and $\bar{c}$ is the average concentration of retained solute in the solution (gm/cc).

This results in a buildup of concentrated solute at the membrane surface, which then back diffuses into the bulk of the solution under the influence of the concentration gradient built up and molecular and eddy diffusion within the fluid above the membrane. This back transport rate is given by the equation:

$$J_{p_{from}} = K_h(c_{mem} - \bar{c}) \tag{2}$$

where: K_h is an overall mass transfer coefficient, (cm/min), which describes the movement of protein away from the surface by molecular and convective diffusion; and c_{mem} is the solute concentration at the membrane surface.

Initially, the rate of convective transport of solute to the membrane is larger than the back transport rate. This results in a further concentration of solute at the surface, until a sufficiently high concentration is reached where the solute cannot be further concentrated due to restricted mobility of solute molecules in the surface layer. Whether this polarized layer is a gel rather than a liquid is open to speculation; perhaps it is sufficient to say the solute molecules are approaching a "close packed" configuration. This concentration has been designated C_g. Once C_g is reached, further buildup of the solute layer must occur by thickening, rather than by further increase in concentration. A concentration profile such as that shown in Figure 4 results.

The layer of concentrated solute at the membrane surface acts as another membrane in series with the original membrane, and retards the flow of ultrafiltrate. Eventually, the ultrafiltration rate is retarded to the point where the convective transport of solute to the membrane surface can be balanced by the back transport rate, and no further buildup of solute occurs. This balance may be expressed by the equation:

$$J_w \bar{c} = K_h(C_g - \bar{c}) \tag{3}$$

This steady state is evidently reached in less than a minute, as shown by invariance of ultrafiltration rates with time (illustrated in Figure 5).

Before application of this model to specific flow geometries is discussed, a number of fairly subtle implications resulting from it should be noted. In the first place, the final ultrafiltration rate at steady state is shown to be controlled by the rate at which solute can be transferred <u>from</u> the membrane surface <u>back</u> into the bulk fluid. Thus, operational variables which aid in back transport of concentrated solute from the membrane (such as more rapid stirring, higher shear rates at the membrane surface, or increased temperature to aid molecular diffusion) will directly increase the final ultrafiltration rate.

Note, however, that variables which simply increase initial ultrafiltration rates without providing a compensating mechanism to increase the rate of back transport of solute will not result in an increased steady state ultrafiltration rate. Thus, an increase in transmembrane pressure drop, which provides an increased driving force for ultrafiltration but does not aid back transport, would simply result in the buildup of a <u>thicker</u> or <u>denser</u> solute cake. The steady state ultrafiltration rate would be reduced to its initial value, and Equation (3) would be once again obeyed. This explains the invariance of ultrafiltration rates with pressure observed in much ultrafiltration data at higher pressures. (Figures 6 and 7 show examples of this pressure independence for human blood plasma and hemicellulose.)

Similarly, the use of a more open membrane with lower resistance to ultrafiltrate flow would not result in a net increase in the final ultrafiltration rate; the solute layer would, again, simply grow thicker, reducing the steady state ultrafiltration value to where it could be balanced by the back transport rate. The invariance of ultrafiltration rates with membrane properties is shown in Figure 8 with data again obtained from the ultrafiltration of whole human plasma.

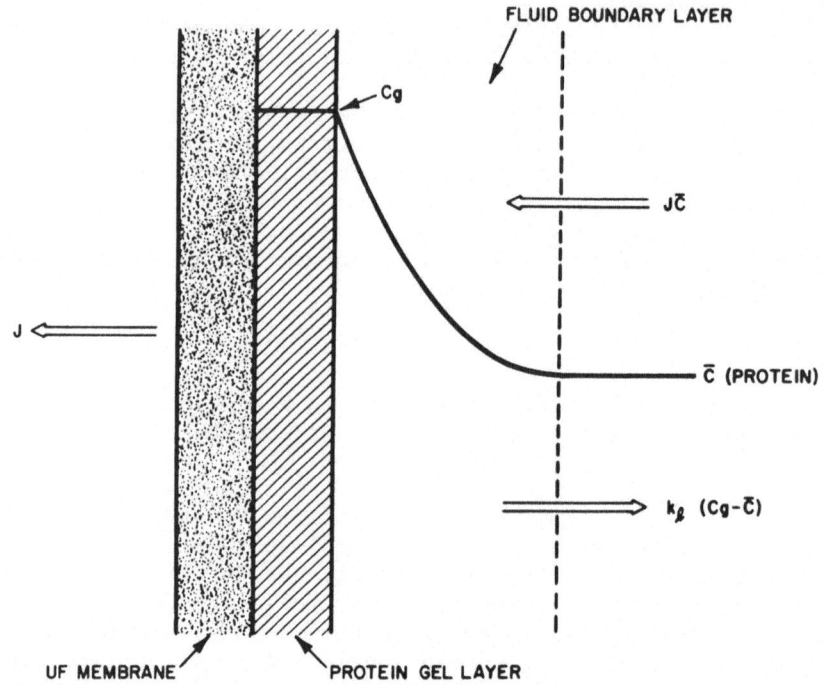

Fig. 4. Physics of the concentration polarization process.

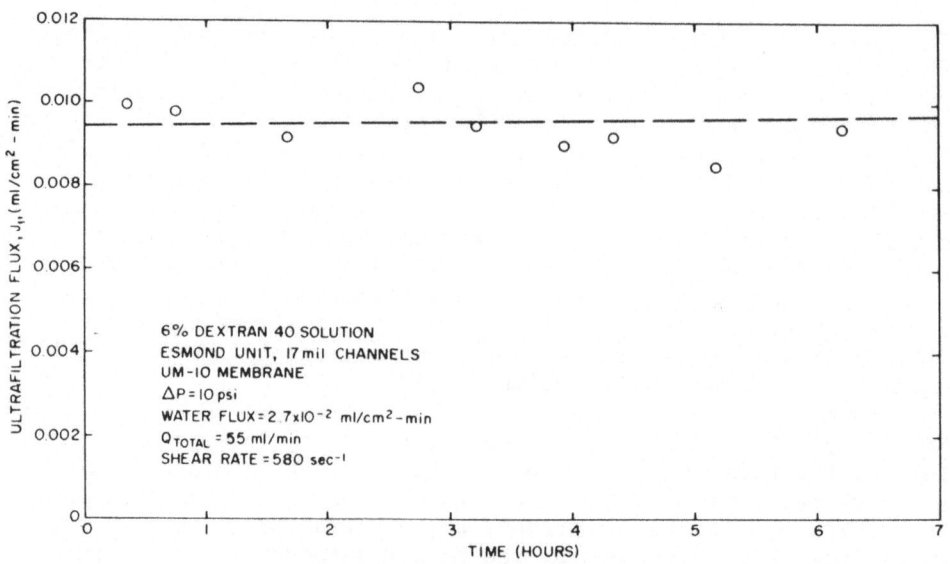

Fig. 5. Invariance of ultrafiltration rates with time.

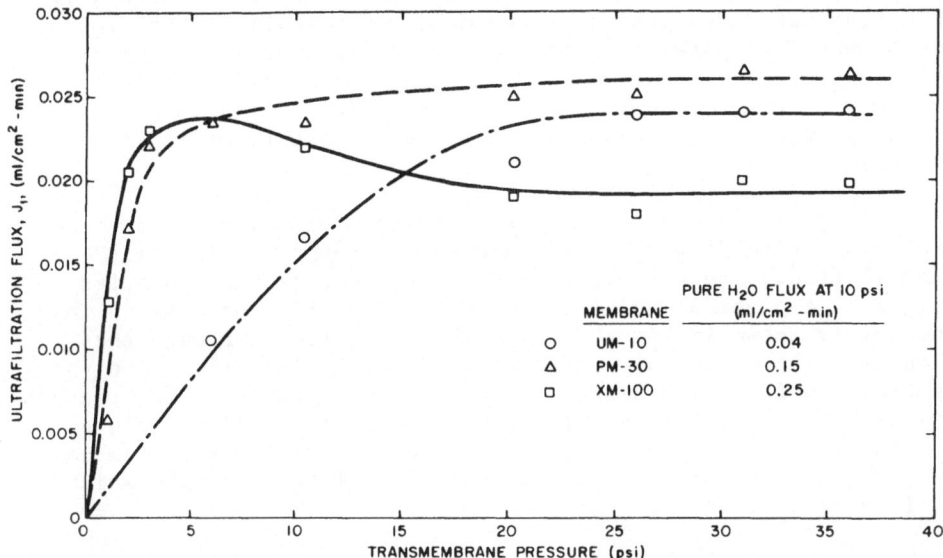

Fig. 6. Invariance of ultrafiltration rate with pressure.

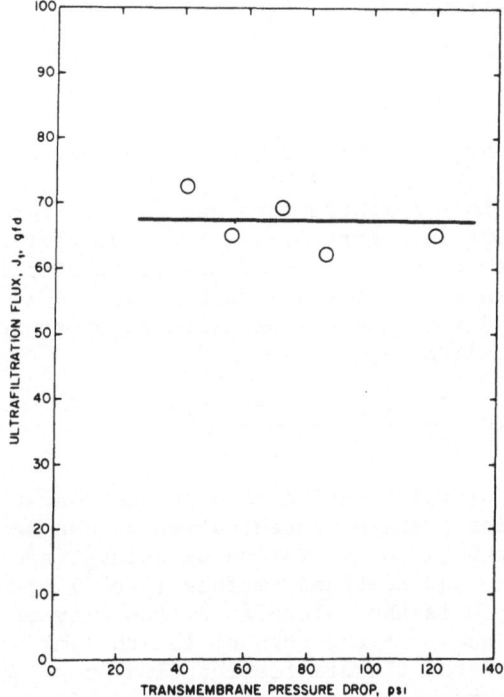

Fig. 7. Invariance of ultrafiltration rate with pressure.

The model discussed above allows the development of quantitative
relationships describing the dependence of ultrafiltration rates
on the physical parameters of the ultrafiltration system. These
relationships quantitatively predict the ultrafiltration rate, J,
in terms of the concentration of the solution, the shape and size
of the flow channels, channel length, and velocity of fluid through
the channels.

We will use as an example a current design for polarization control - solution flowing in laminar flow through a thin "slit"
channel (i.e., one with width and length much greater than its
height). A physical model of the ultrafiltration process in this
channel is shown in Figure 9. As the fluid flows through the
channel, a boundary layer of deposited solute is built up on the
membrane surface. We assume that the solute concentration reaches
C_g very shortly after entering the channel, and further solute
buildup occurs through thickening of the solute layer of concentration, C_g. We wish to predict the ultrafiltration rate which
will occur in this channel.

We have shown, however, that the ultrafiltration rate is a dependent
variable, controlled by the back transport of solute from the membrane surface into the flowing solution. This relationship may be
expressed by rearrangement of Equation (3):

$$J = K_h \left(\frac{C_g - \bar{c}}{\bar{c}} \right) \tag{4}$$

The problem is therefore reduced to solving for K_h.

K_h may be found in straightforward engineering terms by modeling
the back transport in Figure 4 as a mass transfer from a permeable
wall of constant concentration. Exact mathematical solutions for
this model (known as the Levéque solutions) are available for
slit channels and for channels of circular cross section, and are
given by the equation:

$$J = 49 \ln \frac{\bar{c}}{C_g} \left[(\dot{\gamma})_{y=0} \frac{D^2}{L} \right]^{1/3} \tag{5}$$

where: J is the ultrafiltration rate per unit area (ml/cm^2-min);
C_g is the limiting protein concentration at the membrane surface
(g/cc); $\bar{c}$ is the bulk concentration of solute (g/cc); $(\dot{\gamma})_{y=0}$ is
the shear rate at the membrane surface (sec^{-1}) (y = distance from
membrane in cm); D is the molecular diffusivity of retained protein (cm^2/sec); and L is the channel length (cm). (The resulting
relationships for the ultrafiltration flux in terms of average
fluid velocity through the channel are given for slit and circular cross section channels in Table IV.)

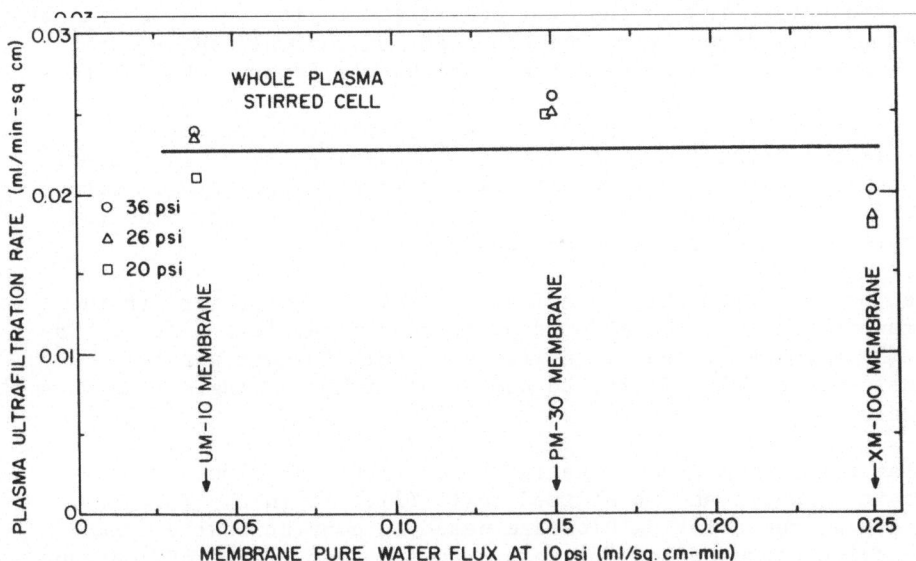

Fig. 8.   Invariance of ultrafiltration rate with membrane properties.

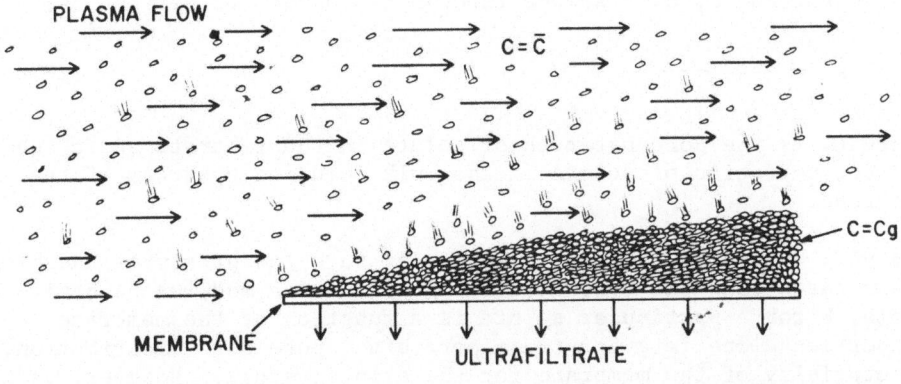

Fig. 9. Development of concentration polarization boundary layer.

Thus, the water flux in the "gel polarized" region would be expected to vary directly as the cube root of the wall shear rate
and inversely as the cube root of the channel length. Data fit-
ting these correlations is shown in Figure 10. The flux would
also be expected to vary proportionately with the logarithm of
the bulk concentration, with a plot of J versus log $\bar{c}$, intercept-
ing the log $\bar{c}$ axis at a concentration corresponding to C_g. Data
plotted in this fashion are shown in Figures 11 and 12. The in-
tercepts appear physically reasonable.

It can be seen from this relationship that the shear rate at the
membrane surface is the major depolarizing parameter. Also, Equation (3) shows that operation with a number of short parallel
channels is to be preferred to operation with one single long
channel.

Similar relationships are available for turbulent flow of solutions in channels and in stirred cells (2). As in the laminar
flow cases, the relationships are based on conventional chemical
engineering correlations for mass transfer from a wall of constant
concentration. The major difference in the turbulent flow cases
is that the relationships are based on empirical correlations
rather than exact mathematical solutions (as in the laminar flow
case), since the solutions are not available. In general, for
turbulent flow, ultrafiltration rate varies to a greater degree
with fluid shear across the membrane surface than it does in
laminar flow (the exponential dependence is closer to 2/3 than to
the 1/3 dependence in laminar flow). Figure 13 shows the turbulent region data for the ultrafiltration of bovine serum.

The rejection, R, of a solute through a membrane is defined as:

$$R = 1 - \frac{C_f}{\bar{c}} \tag{8}$$

where C_f is the concentration of solute in the filtrate and c is
the concentration of solute in the bulk solution upstream of the
membrane.

For very dilute solutions, operating at such low pressures and high
shear rates that the concentration polarization modulus is negligible, R for a particular solute is a function of the membrane
properties alone (e.g., average pore size, pore size distribution,
absorptivity of the membrane for the solute, etc.). However, as
the operating conditions are changed, and concentration polarization becomes significant, the effect of the gel layer formed by
this polarization becomes equivalent to the interposition between
the primary membrane and solution of a secondary membrane through
which all ultrafiltrate must pass. Clearly, if the solute retention characteristics of this secondary membrane are different from
those of the primary membrane, this fact will be reflected in the
composition of the ultrafiltrate.

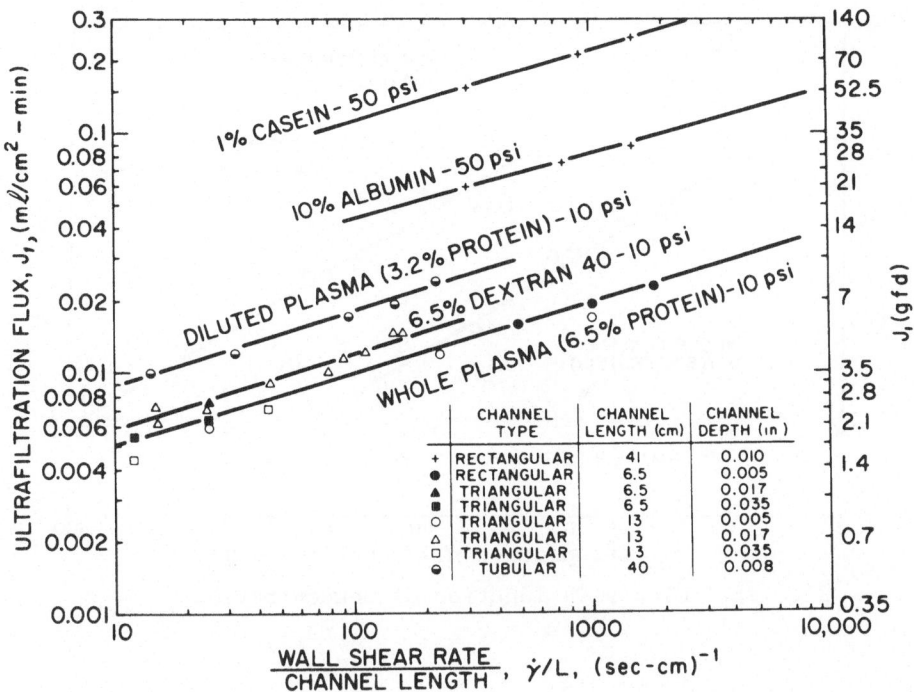

Fig. 10.   Flux as a function of wall shear rate and channel length.

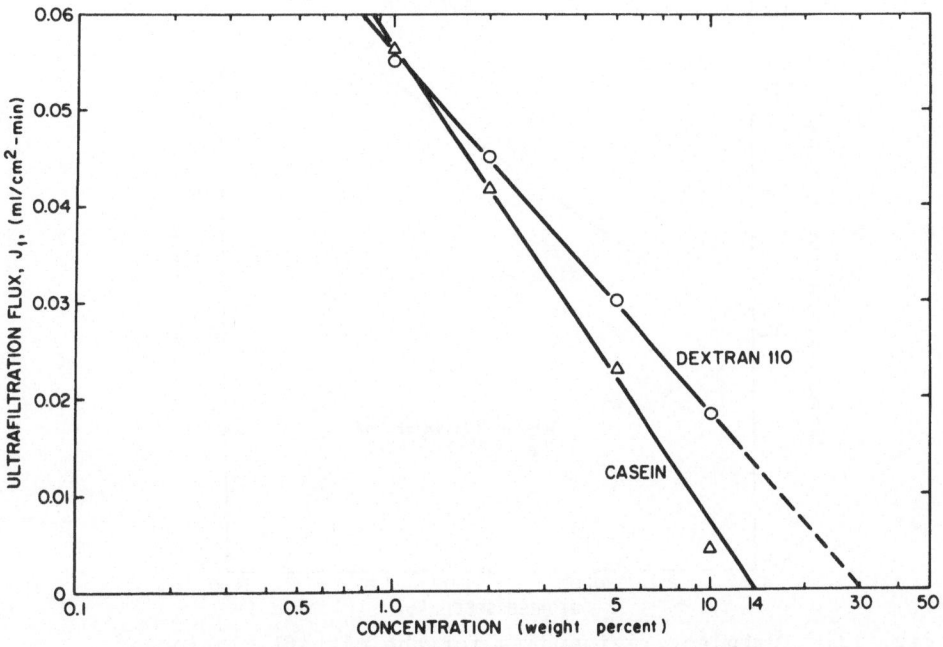

Fig. 11.   Flux as a function of concentration.

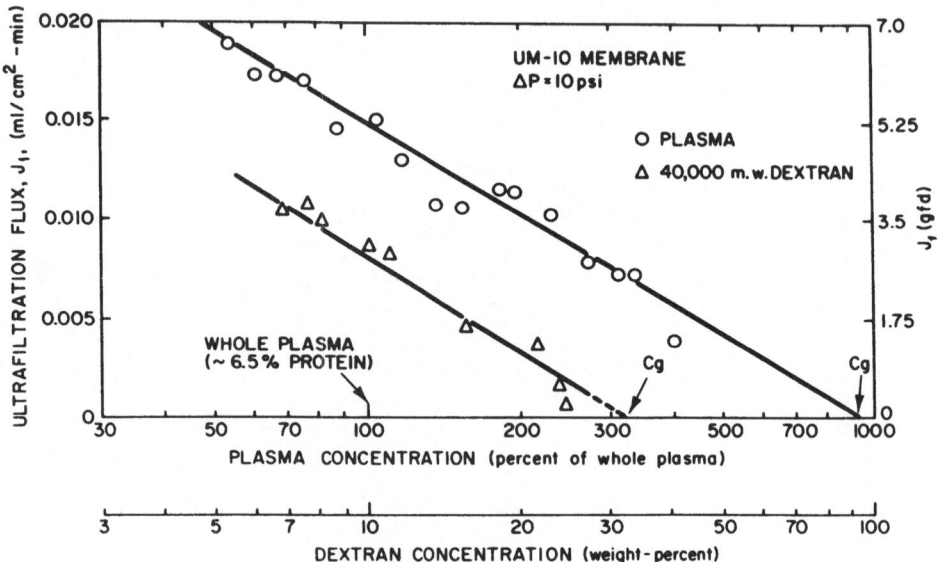

Fig. 12. Flux as a function of concentration.

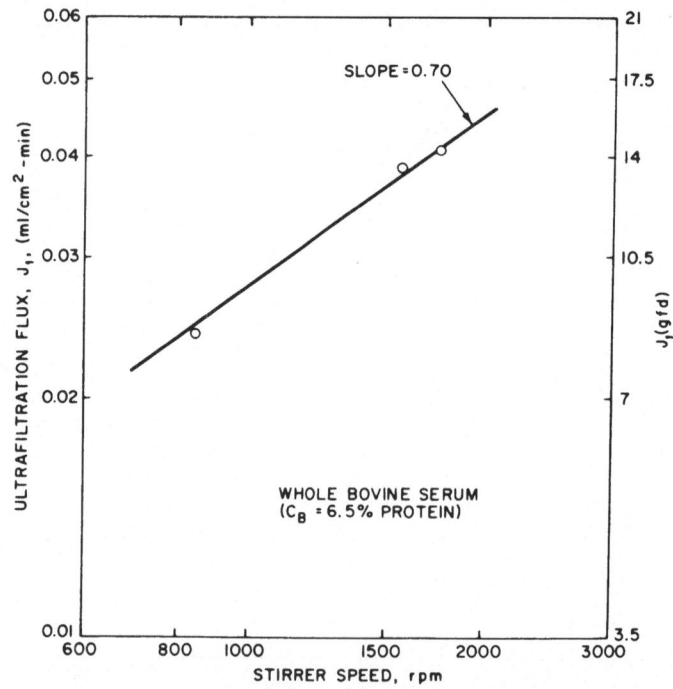

Fig. 13. Turbulent region data for the ultrafiltration of
bovine serum.

If the feed solution contains a single solute, and if the primary
membrane is essentially completely impermeable to that solute,
the formation of a gel layer of solute (while it may markedly reduce flux) will not influence solute retention; the ultrafiltrate
will be solute free with or without polarization. On the other
hand, if the primary membrane is partially permeable to the solute, the effects of polarization on solute retention will be a
function of the mechanical properties of the gel layer. If the
layer is viscous but still fluid, so that solute molecules within
the layer are relatively mobile, then the locally high solute concentration near the membrane surface will cause an increase in
the solute flux through the primary membrane, with attendant decrease in rejection. For example, let us assume that the solute
is of uniform molecular weight, and is non-aggregating (i.e., sol-
ute-solute interaction is negligible) and that the gel layer remains mobile. (This situation is likely to obtain when ultrafiltration is conducted at very low pressures, with quite dilute solutions, and/or with solutes which do not form gels except at
quite high solids concentration.) Let us further assume that the
finite rejection coefficient is caused by a distribution of pore
sizes in the membrane, some of which are large enough to pass
solute, and others of which are too small for solute passage.

If a fraction, α, of the total solvent flux passes through the
solute rejecting pores (while a fraction, $1 - \alpha$, carries solute
at a concentration, C, through the membrane), the reflection coefficient at infinite dilution (or at a concentration polarization modulus, $C_w/\bar{c}$, of 1.0) is given by:

$$R = 1 - \frac{(1 - \alpha)\, \bar{c}}{\bar{c}} \qquad\qquad (9)$$

However, as the concentration at the membrane surface is increased
above the bulk concentration by the concentration polarization
mechanism, the solute which is carried through the membrane is
then at a higher concentration, C_{pore} (where $C_w \gtrsim C_{pore} > C_{bulk}$
and C_{pore} increases with increasing C_w). R is then given by:

$$R = 1 - \frac{(1 - \alpha)\, C_{pore}}{\bar{c}} \qquad\qquad (10)$$

Thus R decreases as the concentration polarization modulus (and
therefore C_w) increases. This effect is illustrated in Figure 14
which presents data from ultrafiltration experiments on dextran
solutions of various molecular weights in a stirred cell (3).
The rejection coefficient can be seen to decrease with increasing
pressure (as would be expected in the pre-gel polarization region,
since increasing pressure increases the concentration polarization modulus).

Similar effects were encountered when defatted milk whey was fractionated using a series of Diaflo ultrafiltration membranes. Defatted milk whey is a complex mixture of proteins of diverse molecular weights, as shown in Table V. Upon fractionation of this
mixture through a cascade of membranes with differing molecular
weight cutoffs, it was found that the lowest molecular weight
fractions (< 10,000) of the mixture passed through membranes with
higher cutoffs, unimpeded by the polarization layer on the membrane surfaces. The medium molecular weight protein fractions,
however, were greatly hindered by the polarized layer, and were
unable to pass through even the most open membrane. Data from
the run are shown in Tables VI and VII. Comparison of actual and
theoretical yields in Table VII shows that Protein Fractions C,
D, and E, which have been expected to pass through the XM-50 membranes, were instead largely retained by the membrane, presumably
because of the polarized layer formed above this membrane by the
higher molecular species of Fractions A and B.

Conversely to the phenomenon discussed above, if the polarized
layer is coherent and truly gel-like, so that molecules within
the layer are essentially immobile (a situation encountered at
high pressure, with concentrated solutions, and with solutes
which are strong gel formers) it becomes virtually impossible for
solute molecules to pass through the layer and reach the primary
membrane. In such a case, the gel layer serves as a partial or
complete barrier to passage of solute, and the retention efficiency increases.

The influence of polarization and gel layer formation on retention becomes most marked when the solutions being ultrafiltered
contain more than one solute, and when the primary membrane is
retentive for one or more solutes, yet permeable to others. In
this case, the gel layer formed of the retained solute or solutes,
if it is of very fine pore texture, or if it is composed of molecules which interact strongly with the other solutes in the solution, may be partially or completely retentive for solutes to
which the primary membrane is permeable. Under those circumstances,
the smaller molecules are retained by the gel layer, and the apparent rejection by the membrane for the smaller molecules increases.

In Figure 15 data are presented on the rejection of human serum
albumin ultrafiltered through XM-300 membranes in the presence of
gamma globulin. Although the albumin rejection coefficient was
approximately zero in the absence of γ-globulin, it rapidly increased as γ-globulin (the rejection of which is > 0.95) was added
to the solution. The data in Figure 15 fit the equation:

$$R_{alb} = 2.4 \sqrt{Conc. \ \gamma\text{-globulin}} \tag{11}$$

although there is, at present, no theoretical reason known for
this square root dependence.

TABLE IV
RELATIONSHIP OF ULTRAFILTRATION RATE TO FLOW GEOMETRY

General:
$$J = 49 \ln \frac{\bar{c}}{C_g} \left[(\dot{\gamma})_{y=0} \frac{D^2}{L} \right]^{1/3} \quad (5)$$

Channel Shape:

Slits
$$J = 49 \ln \frac{\bar{c}}{C_g} \left[\frac{6U_{avg}}{h} \frac{D^2}{L} \right]^{1/3} \quad (6)$$

Circular Channels
$$J = 49 \ln \frac{\bar{c}}{C_g} \left[\frac{4U_{avg}}{R} \frac{D^2}{L} \right]^{1/3} \quad (7)$$

Nomenclature:

J	= ultrafiltration rate per unit area (ml/cm^2-min)
$\bar{c}$	= average conc. of plasma protein in solution (g/cc)
C_g	= limiting protein concentration at membrane surface
U_{avg}	= average fluid velocity through channels (cm/sec)
$(\dot{\gamma})_{y=0}$	= shear rate at membrane surface (sec^{-1}) (y = distance from membrane in cm)
D	= molecular diffusivity of retained protein (cm^2/sec)
L	= channel length (cm)
h	= channel height, slit channels (cm)
R	= channel radius, circular channels (cm)

TABLE V

ANALYSIS OF WHEY (DEFATTED) STARTING SOLUTION

Estimation of the molecular weights from the elution patterns
following chromatography on Sephadex G-100
(Method of Whitaker)

Component	V	V/V_O	Molecular Weight	Weight % in Mixture	Identification
A	15.5 (=V_O)	1.00	>100,000	12.5	? Macroglobulin
B	20.5	1.37	70,500	5.1	Albumin
C	27.5	1.83	28,800	26.3	-lactalbumin
D	32.8	2.18	14,600	11.8	-lactalbumin
E	45.5	2.83	~5,000	4.1	Peptones
F	50.8	3.27	~2,000	36.8	Peptones
G	57.0	3.70	<1,000	3.3	Amino acids, etc.

TABLE VI
WHEY ANALYSIS
FRACTIONAL DISTRIBUTION ANALYSIS OF THE MEMBRANE-SEPARATED COMPONENTS

Membrane Compartment	Nominal Membrane Compartment Molecular Weight Range*	Total Protein (mg)	Fractions Retained in Compartment (%)						
			A	B	C	D	E	F	G
> XM-50	> 50,000	786	32.5	4.8	40.7	22.1	----	----	---
PM-10 - XM-50	10,000 - 50,000	39	22.8	2.6	11.6	37.8	3.3	15.4	6.4
UM-2 - PM-10	2,000 - 10,000	180	4.7	---	----	7.1	11.3	70.1	6.8
< UM-2	< 2,000	606	1.6	---	----	----	----	96.7	1.8

* Based on known Diaflo membrane retention characteristics for single solutes in solution

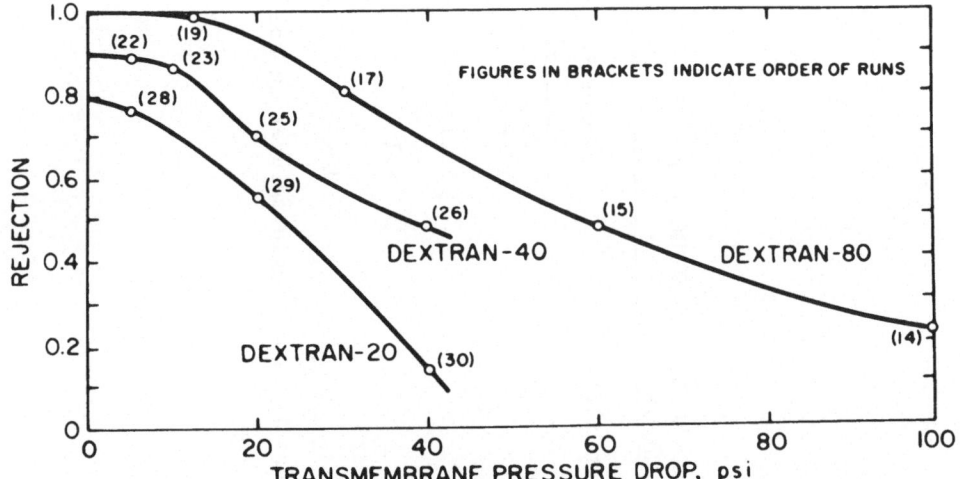

Fig. 14.   Ultrafiltration experiments on Dextran solutions
in a stirred cell.

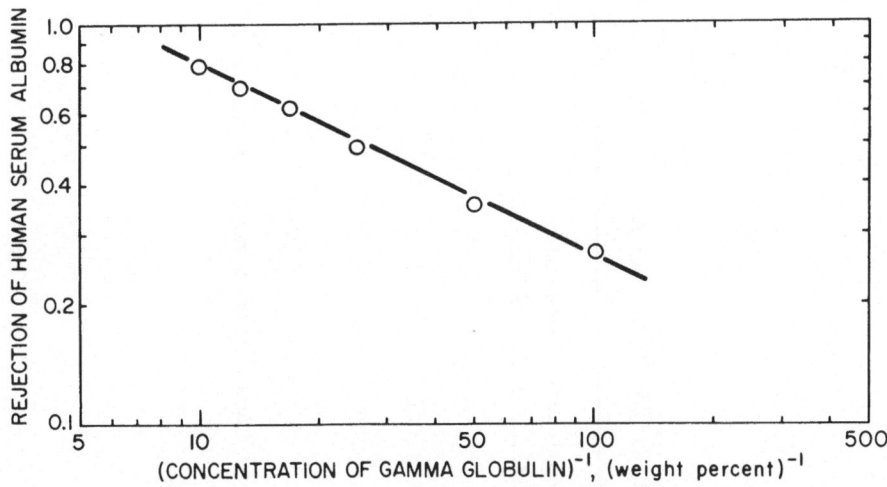

Fig. 15.   Ultrafiltration of human serum albumin in the presence
of gamma globulin.

TABLE VII

COMPARISON OF ACTUAL AND THEORETICAL YIELDS (FROM MOLECULAR WEIGHT) OF TOTAL PROTEIN
ON A PERCENTAGE BASIS

Nominal Membrane Compartment Molecular Weight Range*	Actual Protein Yields (%)	Theoretical Yield (%)
> 50,000	49.7	17.6
10,000 – 50,000	2.4	28.1
2,000 – 10,000	11.2	4.1
< 2,000	27.6	40.1

* Based on known Diaflo membrane retention characteristics for single solutes in solution

Similar data are shown in Figure 16 where the rejection of solutes is shown as a function of molecular weight for single solutes (i.e., the solute is dissolved in a saline solution only)
and for the solutes in a 4% solution of human plasma proteins.

The impact of polarization on solute separation by ultrafiltration is even further aggravated if the membrane impermeable solutes present possess polyelectrolyte character. These properties
are particularly frequently encountered with biological polymers
such as proteins, nucleic acids, and many polysaccharides. In
these cases, the concentrated gel layer formed by polarization may
contain a relatively high ionic charge density, and by virtue of
the Donnan ion distribution equilibrium, excludes even simple
electrolytes. Accordingly it is often observed on ultrafiltration of such polymeric solutes that the ultrafiltrate is significantly depleted in simple electrolytes (salts, acids, bases), despite the fact that the primary membrane (in the absence of pol-   ·
arization) is freely permeable to such substances. While this
phenomenon is undesirable in most purification applications of
ultrafiltration (such as demineralization of protein solutions),
it has effectively been utilized by Kraus and co-workers (4)(5)
for the preparation of "dynamic" reverse osmosis membranes useful
for water desalination.

It is evident that polarization eliminates or greatly reduces the
separative capability of a membrane so that if the purpose of the
ultrafiltration is to effect a separation between larger and
smaller molecules, the efficiency of the process is seriously
compromised by such polarization.

IV.   PROCESS ENGINEERING

Since in many, if not most applications, concentration polarization is the rate limiting phenomenon in ultrafiltration, the design of ultrafiltration equipment and processes is primarily directed toward minimizing polarization within economic limits.
This, in turn, relies upon equipment goemetry and operating conditions which maximize the transport of retained solutes away
from the membrane surface back into the feed solution. The
stirred laboratory cell was the first and simplest way of reducing polarization to levels yielding practical ultrafiltration
rates. An extension of reverse osmosis technology to ultrafiltration equipment design resulted in equipment for "thin channel
ultrafiltration".

This concept is illustrated in Figure 17 where solution flows
through a membrane bounded channel passing solvent through the
membrane and concentrated retentate out the other end of the
channel. High fluid shear rates at the membrane surface decrease
the thickness of the polarized boundary layer. As shown previously,

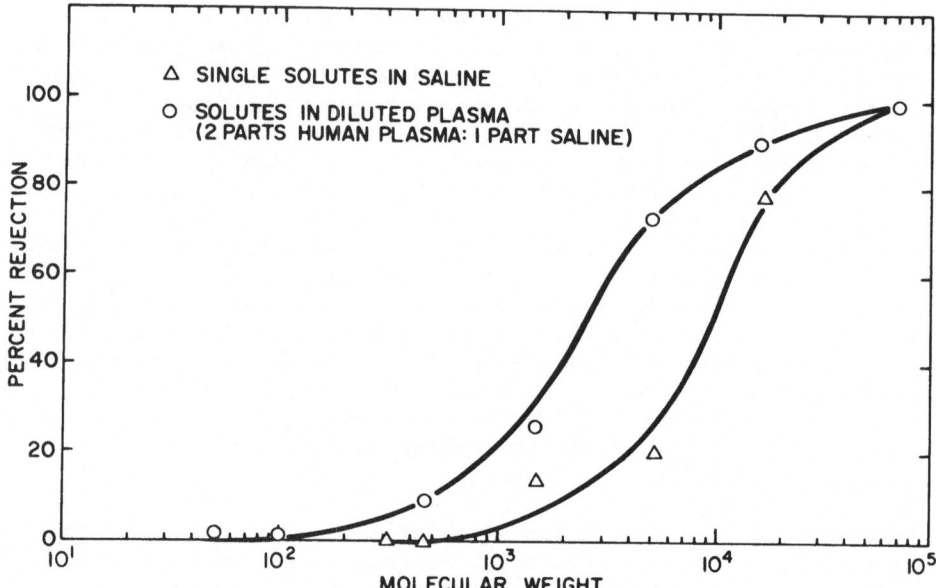

Fig. 16.   Rejection of solutes as a function of molecular weight.

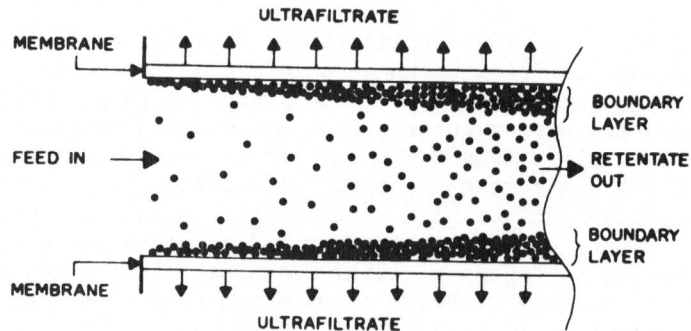

Fig. 17.   Development of the polarized boundary layer.

the shear rate is maximized by increasing the channel velocity
and decreasing the channel height. Obviously, economic optimiza-
tion of channel dimensions and velocities will depend on the nature
of the fluid and must take into account pumping and membrane costs.
Doubling the flux rate through 1,000 square feet of membrane area
at the expense of astronomically high fluid velocities and pressure drops may not be as cheap as adding an additional 1,000
square feet of membrane area.

Compared with other unit operations, ultrafiltration has a great
deal of flexibility with respect to operating modes. Continuous
or batch processing to concentrate, purify, or fractionate can
often be performed on the same equipment with minor in-plant modifications in piping, etc.

Figures 18 through 20 portray three operating modes for continuous or batch concentration.

The recirculation mode of Figure 18 makes possible high channel
velocities regardless of the feed stream flow rate. A fraction
of the concentrate recycle is drawn off as retentate. This system is often used as a batch concentrator without feed or bleed,
in which case the retentate is withdrawn from the reservoir after
appropriate concentration.

Continual recirculation is sometimes deleterious to shear sensitive proteins or polymer dispersions (6), particularly in the
turbulent flow regime (7). Labile solutions dictate operation in
the laminar flow regime, and when extra gentle processing is desired, in the single pass operating mode (see Figure 19). Here,
the solution is dewatered or concentrated to the desired level in
one pass through the ultrafiltration module. Although a pump may
be used to furnish the driving force through the module, it may
also be avoided in batch operation with the use of compressed gas.

The recirculating and single pass operating modes discussed above
may also function as single stages in a multi-stage operation.
Figure 20 combines seven single pass operations in a cascade multi-stage separation scheme for complete separation of partially
rejected solutes from their solvent phase.

Purification from low molecular weight contaminants is partially
effected during concentration since the water removed contains
some of these contaminants. Purification of solutes at constant
concentration can also be achieved with ultrafiltration. Figure
21 features a batch diafiltration operation for removal of low
molecular weight contaminants. As ultrafiltrate, containing lower molecular weight materials, is removed through the membrane,
dialysate (buffer or deionized water) flows into the solution
reservoir at the same rate. The matching of dialysate flow rate

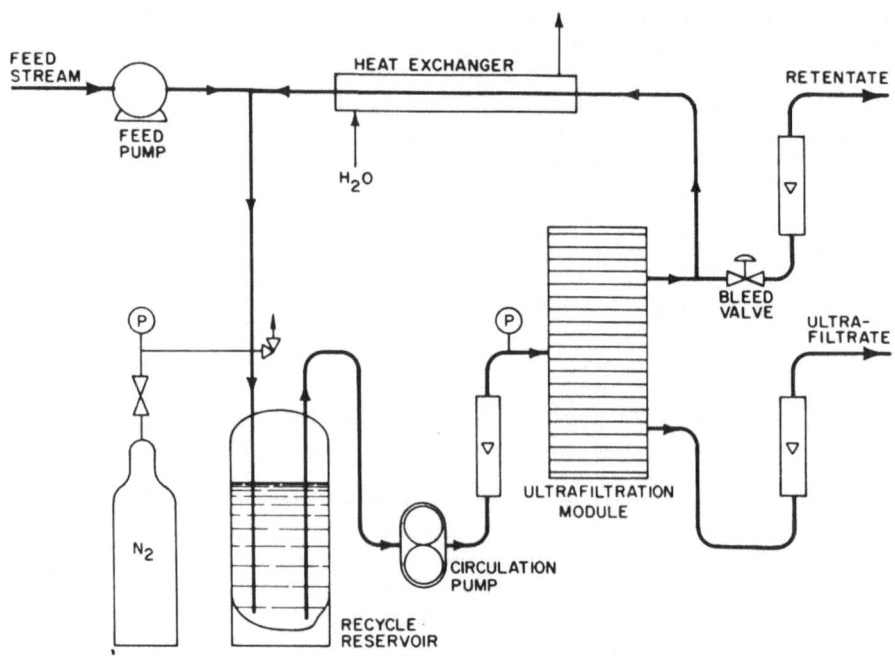

Fig. 18.   Recirculating operating mode.

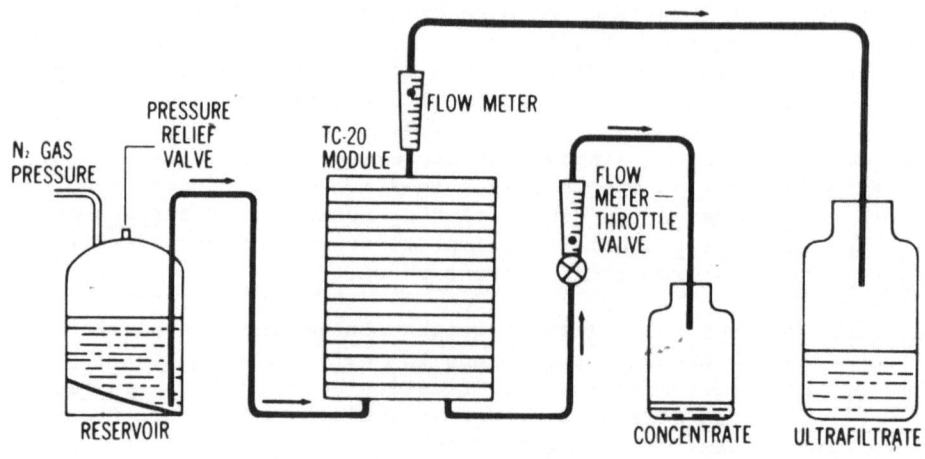

Fig. 19.   Single pass operating mode.

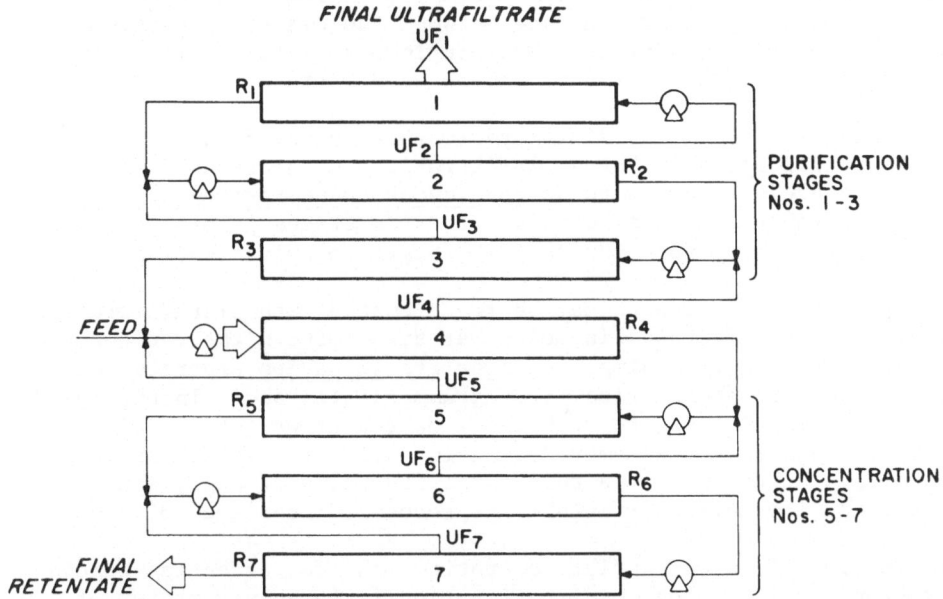

Fig. 20. Cascade multistage separation.

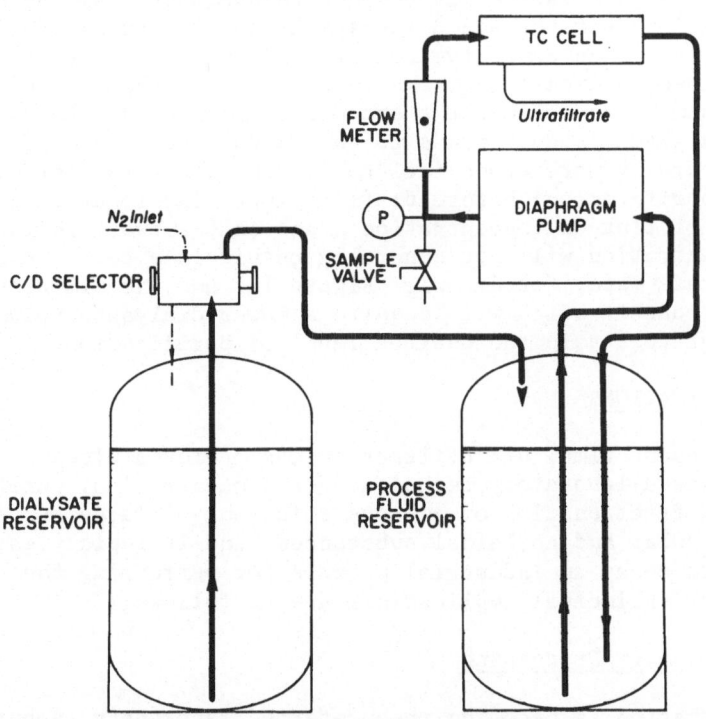

Fig. 21. Batch diafiltration.

with ultrafiltrate flux rate is normally achieved in batch processing with a pressure balance between the dialysate reservoir and the solution reservoir. For continuous diafiltration, level controllers in the reservoirs may be used instead.

The excessive volume of dialysate (wash solvent) required for complete purification in batch diafiltration can be reduced with a staged process referred to as <u>continuous countercurrent diafiltration</u> in Figure 22. The addition of extra stages permits a lower wash solvent flow rate for the same degree of purification.

Multi-component mixtures may be fractionated with ultrafiltration provided the difference in molecular size between the various components is large. A simple <u>cascade fractionation</u> apparatus employed in our laboratories is diagrammatically shown in Figure 23. Here, solvent is used to wash lower molecular weight materials through each of three membranes having different molecular weight cutoffs. At the end of a run, three fractions of high, intermediate, and low molecular weight are recovered from each of the cells.

The choice of an appropriate operating mode for effecting concentration, purification, or fractionation is largely a matter of economics. For example, continuous countercurrent diafiltration requires a higher capital investment than single stage or batch diafiltration, but the added investment may be warranted if the wash solvent is an expensive buffer. Likewise, if several stages are required in processing, the order in which these stages are performed may also affect the economics materially. For example, suppose a given product needs to be concentrated and diafiltered; it is generally more economical to concentrate first (effecting partial purification) before diafiltering. The logarithmic dependence of flux on concentration means that a two-fold increase in the concentration will not generally result in a two-fold decrease in flux rate (see, for example, Figure 11) and the diminished volume from concentration will require a lower dialysate volume throughout to attain the desired level of purification.

V.   APPLICATIONS

At the present time, ultrafiltration has achieved widespread acceptance as a laboratory technique for concentration, purification, and fractionation of aqueous solutions of large numbers of macromolecular and colloidal substances, and is rapidly gaining in importance as an industrial process for performing these same functions. Principal applications are as follows.

A.   Laboratory Processing

Ultrafiltration is becoming the preferred laboratory technique for concentrating dilute solutions of proteins, nucleic acids, enzymes, and other labile or unstable biopolymers, for reasons of speed,

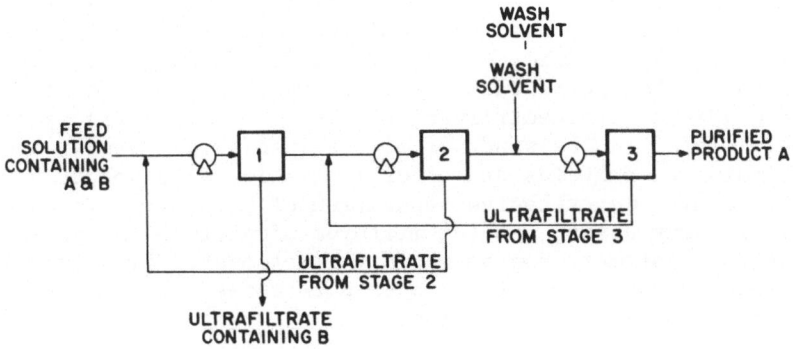

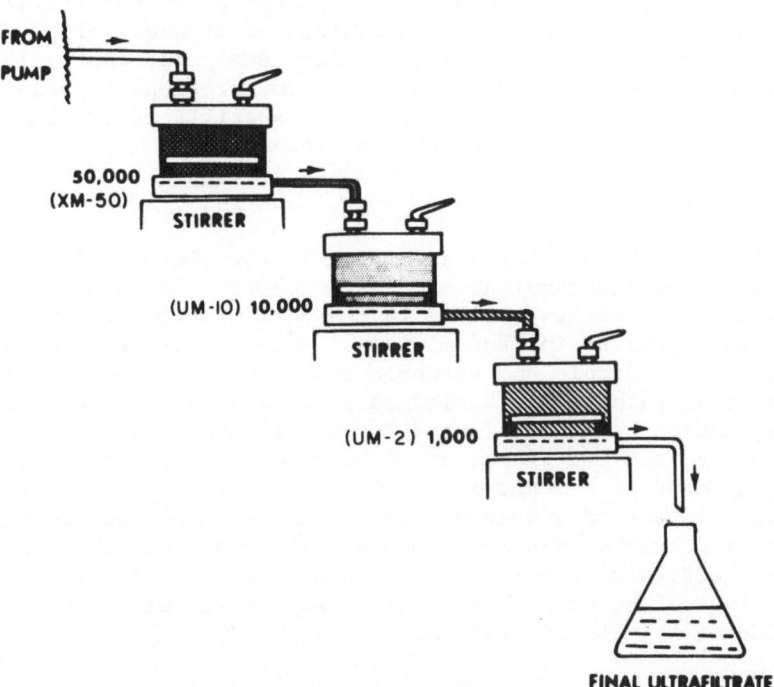

Fig. 23.   Simple cascade fractionation.

convenience, economy, and high product yield without deterioration
or denaturation. A device is now available for the continuous ul-
trafiltrative concentration of gel permeation chromatographic column effluents, permitting more accurate GPC analysis and markedly
improved preparative column yields. Ultrafiltration and diafiltration are now widely used for demineralization and other microsolute (e.g., sugar, urea, etc.) removal from biological prepara-
tions, and for the purification and recovery of biopolymers from
density gradient ultracentrifuge solutions. The process is also
extremely useful in the study of microsolute (e.g., drugs, specific
ions) binding to proteins and other macrosolutes in solution,
since it permits quantitative separation of bound from unbound
solute. In many cases, complex mixtures of proteins or other
water soluble polymers can be rapidly fractionated by ultrafiltration through membranes of the proper pore sizes, into fractions of
nearly homogeneous molecular size.

Medical and clinical diagnostic laboratory uses for ultrafiltration have also been evolving at an increasing rate. These include
the recovery and isolation of trace concentrations of proteins
from biological fluids such as urine and cerebrospinal fluid; the
rapid (and non-destructive) deproteinization of whole blood, plasma, and serum preparatory to wet chemical analysis; the analysis
of trace biologicals in body fluids by immunological and radioim-
mune assay methods; and the analysis for specific macrosolutes in
body fluids by enzymatic degradation procedures.

B.   Industrial Processing

Ultrafiltration is now being exploited in the biological, pharmaceutical, food, and chemical process industries for a host of important separations, as well as in the purification of water and
renovation of industrial and municipal aqueous wastes. The process provides a simple and economic means for both concentrating
and purifying such labile biologicals as viruses, vaccines, and
enzymes, and is being considered for use in the concentration,
purification, and fractionation of human plasma proteins. It is
also being used as a means for isolating and recovering microbio-
logically elaborated substances (e.g., antibiotics, hormones) from
fermentation broths, and for removing allergens and other immuno-
logically active macrosolutes from such products. Sterile, virus
free, pyrogen free water for medical and pharmaceutical use is being produced by ultrafiltration.

In the food industry, ultrafiltration is receiving intensive evaluation and development as a process for recovering concentrated,
purified protein for human and animal use from such by-product
streams as cheese whey, animal blood, and fish stick water. The
economics of large scale ultrafiltration are sufficiently favorable that such products can be profitably manufactured, while at

the same time a substantial pollution problem is eliminated. Ultrafiltration is also emerging as a formidable competitor to alternative dewatering processes such as evaporation, vacuum drying, and freeze drying for such food products as egg white, whole and skim milk, gelatin, vegetable gums, and the like. Enzymatic food conversion processes such as starch hydrolysis to sugar, proteins to amino acids, etc., are both facilitated and improved by ultrafiltration by permitting quantitative recovery of the hydrolysates without contamination by enzyme or residual undegraded materials.

In the chemical industry, ultrafiltration is being used to concentrate and demineralize polymer latices and emulsions; a particularly rapidly growing application in this area is the processing of electrocoating paint primers used in the automotive and appliance industries. In electrocoating systems, ultrafiltration is used both to remove selectively from the electrocoating bath the electrolyte impurities which accumulate during electrodeposition, and to recover colloidally dispersed paint from the rinsing stages following the coating operation. Other emerging industrial chemical applications include the concentration and purification of aqueous pigment dispersions, colloidal silica suspensions, and the like.

In water treatment and water pollution abatement, ultrafiltration is promising to become an important auxiliary to or substitute for other more conventional water treatment processes. Direct ultrafiltration of contaminated water sources yields an ultrafiltrate free of microorganisms, virus, mineral colloids, and organic macrosolutes, frequently yielding a potable product from an impotable supply. Microbiological water treatment processes (e.g., activated sludge treatment) are being improved and beneficiated by ultrafiltration, which allows quantitative removal of the microbiology from the water without loss in microorganism viability. Interest is also growing in the use of ultrafiltration as an adjunct to distillation, reverse osmosis, or electrodialysis for water desalination; ultrafiltration of the feed water to the desalination units removes colloidal and macromolecular organic and inorganic substances which contribute to fouling of heat transfer surfaces and membranes.

Lastly, a rather important medical application for ultrafiltration is now in advanced development, which may have important consequences in the treatment of chronic kidney disease. This is the so-called "Diaphron" or hemodiafilter, which is a device for ultrafiltering whole blood to remove metabolic wastes such as urea and uric acid without loss of blood proteins or formed elements. This technique appears to have numerous advantages over conventional hemodialysis, and preliminary clinical trials are now in progress.

In overview, ultrafiltration is now evolving as one of the most important novel molecular separation processes to have been developed in the past twenty years, and its potential for use in the laboratory and in industry is yet far from being fully exploited. The next decade should see a substantial expansion in utilization of the process, along with the development of considerably more sophisticated membranes, devices, and systems than are available today.

LITERATURE REFERENCES

(1)   Michaels, A. S., "Ultrafiltration", from Advances in Separations and Purifications, edited by E. S. Perry, 1, Intersci-ence Publishers, New York (1968).

(2)   Blatt, W. F., A. N. Dravid, A. S. Michaels, and L. Nelsen, "Solute Polarization and Cake Formation in Membrane Ultrafiltration: Causes, Consequences, and Control Techniques", from Membrane Science and Technology, edited by J. E. Flinn, Plenum Press, New York (1970).

(3)   Baker, R. W., J. Appl. Poly. Sci., 13, 369-376 (1966).

(4)   Shor, A. J., K. A. Kraus, J. S. Johnson, and W. T. Smith, I & E C Fund., 7 (1), 44-48 (1968).

(5)   Johnson, J. S., K. A. Kraus, et al, French Patent 1,497,295.

(6)   Charm, S. E. and B. L. Wong, "Enzyme Inactivation with Shearing" (in manuscript).

(7)   Charm, S. E. and J. Lai, "A Comparison of Ultrafiltration Systems for Biologicals" (in manuscript).

ELECTROKINETIC MEMBRANE PROCESSES

Milan Bier

Veterans Administration Hospital and University of Arizona

Tucson, Arizona

INTRODUCTION

All membrane processes have the common objectives of fractionation, concentration, and/or purification. The driving forces may be concentration or pressure gradients, as, for example, in dialysis, reverse osmosis, or ultrafiltration. In the present paper we wish to discuss a family of interrelated processes wherein the same objectives are achieved through electrical driving forces. Most important among these is electrodialysis using ion permselective membranes, but we shall omit it from our considerations, its applications being widely known and studied (1). Instead, we shall focus on processes applicable primarily to:

    Large scale electrophoretic fractionation and concentration.

    Non-clogging electrofiltration of particulate suspensions.

    Electroosmotic concentration and desalting of macromolecular systems.

    Electroadsorption at the membranes in absence of the electroosmotic effect.

These processes developed from our studies of forced flow electrophoresis (FFE), first conceived as a large scale electrophoretic technique, capable of complex biomedical and industrial applications (2-4). In fact, the technique proved applicable not only to electrophoresis, but also electrofiltration (5), electrodialysis (6), electroadsorption (7), and electroosmosis (6, 8). These effects are often overlapping, and clear distinction is impossible. The term FFE is, therefore, retained for electrophoresis alone, and we propose the term electrokinetic membrane processes (EMP) to encompass all electrically driven membrane processes.

Classical colloid chemistry recognizes two electrokinetic phenomena arising from the application of an external electrical field, electrophoresis being defined as the migration of electrically charged ions or particles, and electroosmosis as the migration of the solvent. Complications arise when membranes are interposed in the pathway of the migrating particles, as layers of increased and decreased solute concentration are formed along opposite sides of the membranes, giving rise to cell content polarization. The primary purpose of the present paper is to clarify these various cell phenomena, and relate them to their useful applications.

APPARATUS

The equipment employed consists of a sequence of membranes and filters, kept within a multiframe assembly, the electrical field being imposed by means of external electrodes. Its details have been described repeatedly (1-3,9,10). Depending on the intended use the exact sequence of membranes and filters may vary, as will be shown. The nature of the membranes and filters is not critical, cellulosic dialyzing membranes such as Visking, (Union Carbide Corp.) and a variety of filters, such as filter paper, microporous battery separators, or filters of the type made by Millipore and Gelman Corp. having been used. This is in sharp contrast with other processes such as ultrafiltration and electrodialysis, where the properties of the membranes are of critical importance. The reason for it is that in EMP the main discriminating factors are not the membrane characteristics but the ionization of the solutes and the pH and conductivity of the solution.

Most work has been carried out with frames having either 100 or 1000 cm^2 effective area, and up to 100 cells have been used in parallel. The potentials applied varied from 6 to 50 volts per cm, and current densities of up to 0.05 amps/cm^2 were used. Circulation of liquids through the apparatus was either through metering pumps or by gravity and, in all instances, only very small resistance to flow was encountered. Equipment of both sizes is available from Canalco, Inc., Rockville, Md.

A - ELECTROPHORETIC FRACTIONATION AND CONCENTRATION

A schematic diagram in Figure 1 illustrates the side view of a FFE cell assembly with the simplest possible arrangements of filters and membranes. It indicates also the location of the top and bottom outlets, for the colloid impoverished and concentrated fractions, resp. Other arrangements of filters and membranes are possible, as will become evident.

Simplest to explain is the performance of the apparatus assuming laminar flow through it. This situation, while never truly reached, is approximated at low colloid concentration, causing negligible cell polarization only. Salt concentrations of the order of 0.01 M are sufficient to minimize electroosmotic effects due to the small charges of the filters and membranes, but not high enough to cause excessive Joule heating. Soluble colloids readily pass through the filters, but are retained by the dialyzing membranes.

While maintaining a downward flow of the feed solution through all input compartments, part of the fluid is forced through the filter by gravity flow or appropriate pumping arrangement. The polarity of the electrical field is so chosen as to force the charged colloids to migrate in a direction opposite to that of liquid flow through the filter, as indicated respectively by the broken and solid arrows in Figure 1. In practice most colloids are found to be negatively charged, and this polarity will be assumed throughout the present discussion.

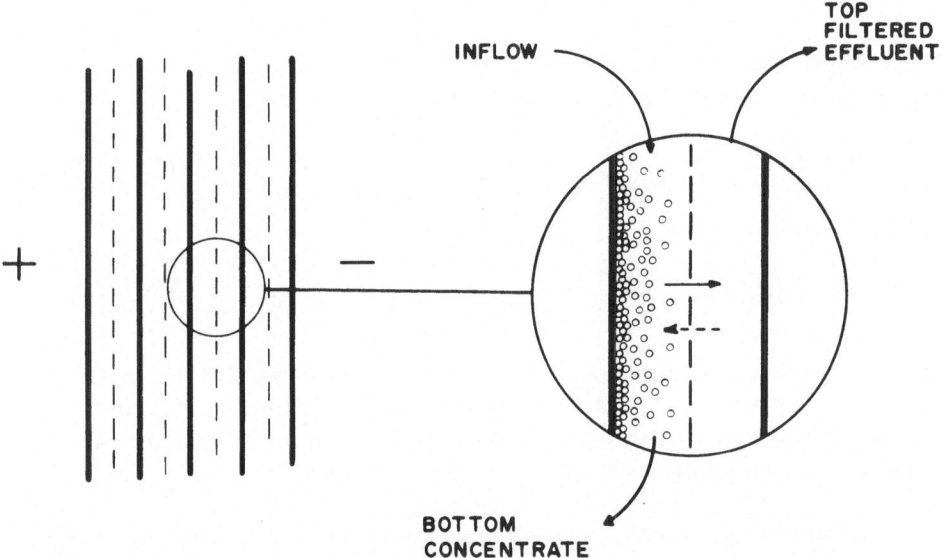

Figure 1: Schematic presentation of a cell pack for FFE. Solid lines - dialyzing membranes; broken lines - filters. Solid arrows - direction of liquid flow; broken arrows - direction of electrophoretic migration.

To those familiar with filter paper electrophoresis, FFE can be presented as a variant in which separation is carried out in a direction perpendicular to the plane of the filter, rather than within the plane as in the conventional procedure. The whole area of the filter is utilized, thus accounting for the high flow rates, but only two fractions are obtainable: the colloid enriched bottom fraction and the colloid impoverished or isoelectric top fraction.

The flow through the filter of a charged colloid species of electrophoretic mobility μ can be expressed in grams per second, $Q_f C_f$, where Q_f is the volumetric rate of flow of the filtered effluent, and C_f its colloid concentration. It is given by the difference between the colloid content of the original feed, and the colloid removed electrophoretically, viz.

$$Q_f C_f = Q_f C_o - \mu \, ENAC_o \tag{1}$$

where C_o is the concentration of the same species in the feed solution, E the voltage gradient (v/cm) applied, N the number of filters used in parallel, and A their effective unit area. By rearrangement, the fraction F of colloid removed is obtained as

$$F = 1 - C_f/C_o = \mu ENA/Q_f. \tag{2}$$

The voltage gradient E can be expressed in terms of amperage, I, and specific conductivity, k, of the solution,

$$E = I/kA \tag{3}$$

and equation (2) becomes

$$F = \mu \, NI/kQ_f. \tag{4}$$

This expression is of interest because it does not comprise any cell dimension factors. Obviously, there is a practical limit to current density, and for many protein separations this was found to be of the order of 0.05 amp/cm^2.

According to the above expressions, concentration of a given charged colloid in the filtered effluent will decrease as a linear function of amperage or voltage applied, until finally 100% removal is obtained. Critical voltage gradient, E_{cr}, (or amperage) can be defined as the minimal value giving total protein removal

$$E_{cr} = Q_f/\mu \, NA. \tag{5}$$

The net effect is as if the electrical field were modifying the properties
of the filters to render them retentive for charged colloid species.

Above equations are at best only a first approximation, as they neglect
a host of other factors such as electrodecantation, electroosmotic effects,
temperature gradients, and polarization, all of which contribute to the
fact that cell content is by no means homogenous. An exact treatment of
these factors would be most difficult, and they will be discussed in only
a qualitative way. It is significant, however, that above equations have
been found to provide adequate guidance in planning of experiments, as
actual data seem to conform to expected results As an example, data on
separation of purified egg albumin from the top effluent are presented in
Figure 2. Several constant flow rates have been employed, and the amperage gradually increased, the effluent concentration being reported as percent of the feed concentration. The solid lines represent curves on the
basis of conductivity of buffer employed and the known electrophoretic
mobility of the protein (11). Experimental points show excellent agreement
with expected values.

The equations can be also used to predict approximate flux rates.
For instance, if in a protein mixture, such as serum or plasma, with a

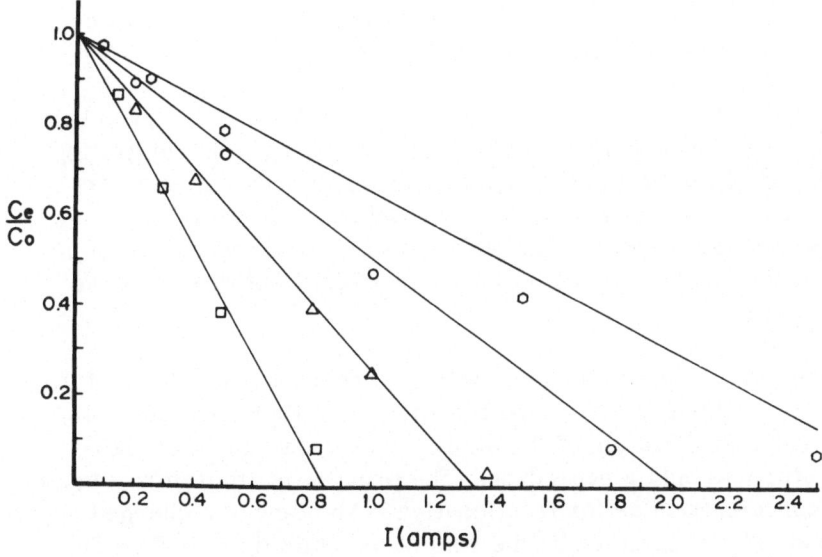

Figure 2: Electrophoretic removal of purified egg albumin from
a phosphate buffer at various flow rates. C_e/C_0 is the ratio of
effluent over original concentration.

conductivity of about $k = 10^{-2}$ mho/cm, complete removal ($F = 1$) is desired for all proteins with a mobility higher than $\mu = 2 \times 10^{-5}$ cm²/volt-sec equation (4) predicts a flux rate of 360 ml/hr per sq ft of filter area, at current density of 0.05 amps/cm². On the other hand, colloids in natural waters or sewage may have a typical mobility of the order of $\mu = 20 \times 10^{-5}$ cm²/volt sec, and if the conductivity $k = 5 \times 10^{-4}$ mho/cm, a flow rate of over 2 gal/hr per sq ft filter area is obtained, at current density of only 0.01 amps/cm². Such flow rates are customarily obtained in actual experiments (5, 9, 12, 13).

As can be seen from above two examples, conductivity of the feed is usually more important than electrophoretic mobility, particularly as the mobility of a colloid usually increases with a decrease in conductivity, all other factors remaining constant. For increasing the fractionation rate of serum, for example, prior desalting through dilution, dialysis, or ion exchange, is advisable. Such desalting increases processing rates manyfold. In a multicomponent mixture, such as frequently encountered in protein systems, for example plasma, fractionation is obtained by virtue of different electrophoretic mobilities characterizing each component. Purification ratio $F_1/F_2 = \mu_1/\mu_2$ for components 1 and 2, is highest if component 2 has zero mobility, i.e. is isoelectric. By their very definition, isoelectric proteins are not affected by an electric field, and will be found in the filtered effluent at highest purity and at their inflowing concentration. Components are rendered isoelectric, of course, through appropriate adjustment of the pH of the solution.

In principle, it appears as if it would be possible to obtain even better fractionation if the two components had opposite charge, one migrating in the direction of liquid flow through the filter, the other in counter-current direction, as assumed above. Frequently, however, protein concentration is sufficient to cause at least some electrodecantation of the filtered species, causing decreased effectiveness of this 'reverse polarity' scheme.

Above apparatus operates essentially as a concentration device for faster proteins, which are removed from the top effluent and added to the bottom one. Purification of the slower component, or of the isoelectric protein, is only a byproduct. The process is, therefore, equally useful for concentration as for fractionation. The flow of a charged species in the bottom effluent is given by the sum of the colloid in the feed and that separated from the filtered effluent,

$$Q_b C_b = Q_b C_o + \mu\, ENAC_o \qquad\qquad (6)$$

and the concentration factor F_c is

$$F_c = C_b/C_o = 1 + \mu \; ENA/Q_b = 1 + FQ_f/Q_b \qquad (7)$$

where the subscript b refers to the bottom effluent.

It may be useful to consider what will be the distribution of components in an idealized mixture of a mobile and an isoelectric component after half the original volume was collected as the top filtered effluent, the other half as the bottom concentrate, assuming a perfect fractionation, F=1, for the mobile, and F=O for the isoelectric component. This can be taken as a unit fractionation step. At this stage, the top fraction will contain the pure isoelectric component at its original concentration, but only 50% of its total amount (half the original volume) will have been recovered. This is schematically illustrated in Figure 3. The bottom concentrate will have all of the mobile component, at twice the original concentration, but will still be contaminated with the remaining half of the isoelectric component. It is completely immaterial, however, how this stage is reached, i.e. whether the two flows were actually identical, or the top flow much slower than the bottom, and the bottom effluent subjected to repeated recycling through the apparatus. Accordingly, essentially identical fractionation of serum proteins was obtained with the ratio of the two flows varying between 0.1 and 10. Nevertheless, to avoid buildup of concentrated protein within the input half of the cells, and to better dissipate the heat generated by the electrical current, the feed solution

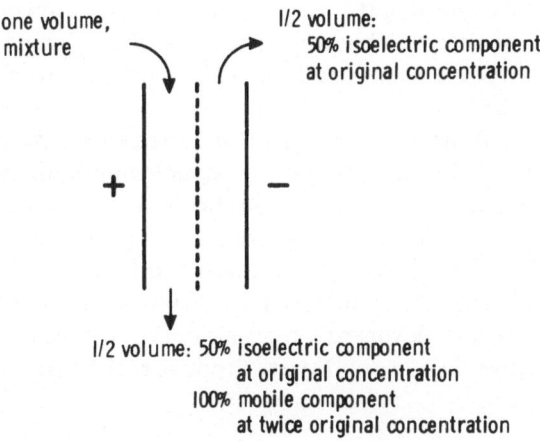

Figure 3: Hypothetical unit processing step in FFE, showing distribution of isoelectric and mobile components assuming equal top and bottom flow and perfect fractionation.

is often rapidly recycled, while the flow of the top effluent is maintained at about 1/5 of the total flow. Only this latter flow rate, of the filtered effluent, needs to be carefully regulated, as it alone controls the quality of fractionation.

During the recycling, the concentration of the faster component keeps on increasing. If this is to be avoided, and to ensure more complete recovery of the slower or isoelectric component, the feed solution can be maintained at constant volume through addition of appropriate buffer. In such a mode of operation, the concentration of the slower component in the top effluent is continuously decreasing.

The nature of the filter was not considered in above equations, but in practice, it was found that the smaller the pore size, the sharper the fractionation. The molecular weight of the colloid is also of consequence. With bacteriophages top effluent concentrations were reduced by as much as 10^7 with respect to inflowing concentrations (7). With average proteins, their molecular weight being an order of magnitude smaller, this decrease is of the order of 10^2. With still smaller polypeptides, the reduction achieved is only of the order 10 (9,14). This interaction between the filter and colloid is probably due to electroosmotic or surface effects within the filter pores.

In above discussion low colloid concentration was assumed giving rise to negligible density gradients only. This is not the usual case. Electrophoretic migration of charged colloids within the cell will result in their accumulation in the immediate neighborhood of the corresponding membranes. According to the well known principles of electrodecantation, as studied by Pauli, Kirkwood and others (2,15-17), this colloid accumulation gives rise to density gradients, causing a significant downward flow of the layer of increased colloid content. FFE utilizes these principles and the downward flow of the liquid through the input compartment is in the direction of electrodecantation. With colloids lighter than water this direction obviously has to be reversed. Additional density gradients arise due to Joule heating. The simple function of the filter described above no longer suffices, as it also functions as a frictional boundary preventing the remixing of this downward flowing layer of increased colloid density with the upward flowing layer, impoverished in colloids, along the opposite membrane.

Nevertheless, FFE is not an electrodecantation method as it does not depend on this effect for the separation of components which is achieved through the flow pattern imposed on the fluid. For one, it has been shown that it applied to very dilute virus suspensions (7), where electrodecantation is ineffective because of insufficient density gradients (17). More important, the flow of the two effluents can be varied at will through appropriate pumping provisions, thus providing FFE with greater flexibility and complete control over operational parameters. In a given system, the quality of fractionation achieved is mainly function of the electric field applied and the flow rate of the filtered effluent, and is relatively insensitive of the downward rate of flow through the input compartment. Fast downward flow provides excellent internal cooling, permitting higher fields for faster fractionation of complex biological materials. Additional cooling, if necessary, can be easily provided by inserting between each two fractionation cells a channel formed by two membranes, through which rapid circulation of a cooled buffer can be maintained. This permits electrodialysis to be superimposed on the fractionation.

We have also assumed that the filters and membranes are electrically neutral and allow unhindered passage of small ions of both polarities. This is essentially true of the membranes and filters employed in absence of colloids. In their presence the accumulation of charged species in the immediate neighborhood of the membranes mimics certain effects of ion exchange membranes (6, 8) and causes a flux of ions across the membranes and into the colloid layer. Paralleling this flux of ions there is also a flux of water across the membranes into the colloid layer. As this water flux is countercurrent to the migration of colloids it effectively prevents their electrodeposition on the membranes. It is this effect which prevents membrane fouling and it is most surprising that its importance has not been recognized by workers in electrodecantation. In absence of strong electroosmotic gradients electrodeposition, or electroadsorption, can occur, as will be discussed separately. The electroosmotic effect can be maximized by using colloids giving highest concentration of charged groups in the neighborhood of the membranes (18), and the use of polyelectrolytes for such purposes will be described in the section on electroosmotic concentration and desalting. As a result of the ion flux the bottom effluent will emerge with increased salt content, while the top effluent is desalted. Surprisingly enough, if the feed is buffered, the pH of the effluents usually remains unchanged, which, however, does not exclude localized pH gradients. In many applications this effect can be neglected, but if rigorous constancy of fractionation conditions is desired, automatic regulation of conductivity of the recycling feed solution may be necessary.

B -ELECTROPHORESIS ACROSS ULTRAFILTRATION MEMBRANES

Electrophoresis as described in the preceding section is primarily applicable to the isolation of the slowest component in a mixture.

Intermediary components can be obtained through cascade subfractionations only, but the more fractionation steps are necessary, the less appealing the method. The fastest component can be also obtained by complete depletion of all slower components, as will be shown later on. It would be desirable, however, if the scheme presented in Figure 1 could be reversed, in order to obtain the fastest component directly. Such a scheme is presented in Figure 4. In this scheme, the polarity of the electric field is reversed; rather than having the fastest component migrate away from the filter, it is made to migrate towards the filter, as indicated by the broken arrow. To retain the slower components, a backwash flow of buffer may be established, as indicated by the shorter solid arrow. The linear rate of flow of the buffer has to be smaller than the rate of electrophoretic migration of the fastest component. This scheme has been tried and found not to be effective, unless it is combined with simultaneous particle size discrimination.

This is possible only under specially favorable conditions, for instance in plasma or serum proteins, where serum albumin is not only the electro-

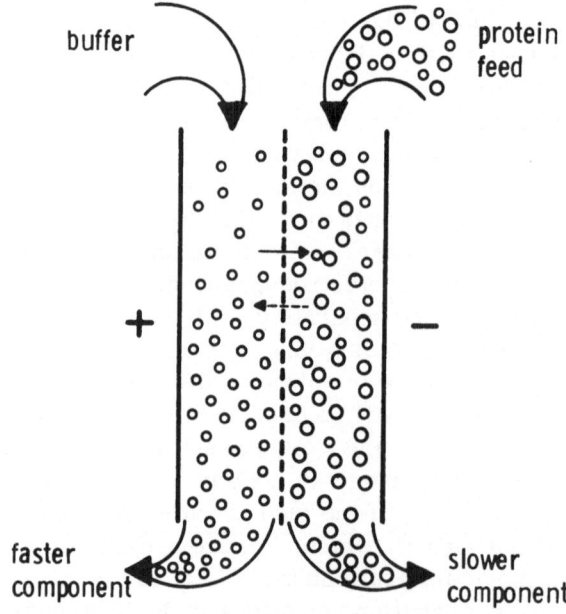

Figure 4: Reversed polarity FFE: broken line indicates ultra-filtration membrane, i.e. a filter which discriminates against some solutes by virtue of their size. Solid arrows - direction of liquid flow; broken arrow - direction of electrophoretic migration.

phoretically fastest major component, but also has the smallest molecular weight. Using modified cellulose membranes of controlled porosity, it has been possible to directly isolate serum albumin, using the scheme illustrated above. Controlled porosity membranes are also available commercially, for instance the Amicon Corp. XM100 membrane with a theoretical cutoff of about 100.000 molecular weight. In the preceding paper of this Symposium, Michaels and coworkers have shown that these membranes, using ultrafiltration, allow passage of albumin, and reject globulins, when each protein is used alone. In mixtures, of the two proteins, the rejected globulins accumulate in front of the membrane, and render it impermeable to albumin as well. This polarization is completely avoided in the above electrophoretic scheme, as serum albumin is forced to electrophoretically migrate across the membrane against a stream of buffer which effectively backwashes the membrane.

Another instance where this procedure has been found useful is in the separation of bacteriophages from bacteria. Millipore Corp. sterilizing filters have been used, which effectively retained all bacteria, while the phages were electrophoretically transported across the filter. Backwash was not necessary, because of the small quantity of bacteria present. The electrical field not only caused the transport of bacteriophages across the filter but also prevented their adsorption on the filter, which occurs in absence of the field. This procedure permitted obtaining bacteria-free bacteriophages for their subsequent culture in absence of unwanted bacterial contamination (7).

C -ELECTROFILTRATION OF PARTICULATE SUSPENSIONS

If instead of filterable colloids, the processed liquid is a suspension of particles large enough to be mechanically retained by the filter, then the top effluent will, of course, always be particle free, i.e., F, the fraction of colloid removed is always equal to 1. In absence of the electrical field and under constant pressure, there will be a more or less rapid decrease in filtration rates due to filter cake deposition. The application of the electrical field causes the particles to migrate away from the filter in the same manner as described above for proteins. It is an empirical observation that most particulate matter in water and sewage carries a large negative charge, and the same is known also for most bacteria, algae, viruses and other microorganisms. The particles are thus prevented from depositing on the filter and, instead, may deposit either on the membrane facing the filter or, preferably, be swept through the bottom opening of the input compartment by the downward stream of concentrated effluent. This is illustrated schematically in Figure 5. Brief periodic

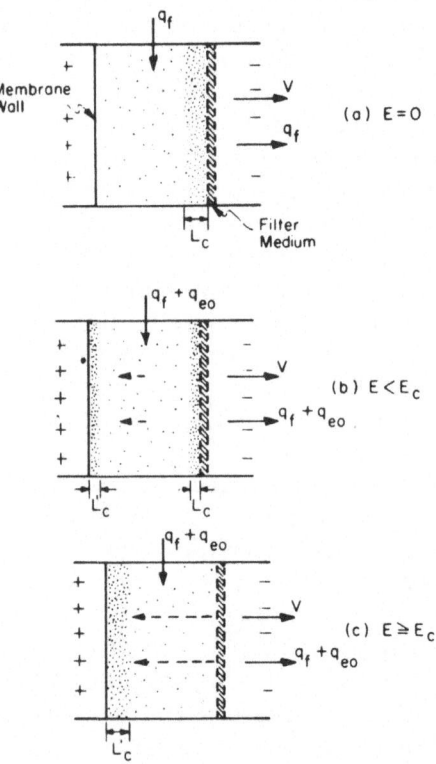

Figure 5: Schematic presentation of three stages in electrofiltration: top - no voltage applied; center - insufficient voltage
applied; bottom - adequate voltage prevents all filter cake deposition on the filter. Deposition on the membrane can be prevented by continuous drainage of the input compartment, and occasion-
al reversals of current polarity. Solid arrows - direction of
liquid flow; broken arrows - direction of electrophoretic migration.

reversals of current polarity may be used to break up any deposits formed
on the membrane. The downward flow of liquid through the input compartment helps to avoid settling problems. With particulate matter such as
oil droplets, which are lighter than the suspending medium, an upward
flow should be used, reversing all directions of in and outflow. On the
other hand, if bottom outflow is prevented, packing of particulate matter
in the input compartment will continue until the latter is completely filled
without significant decrease in flow rates. Tight packing of input compartments without decreased flow rates has been achieved with such diverse
materials as clay, algae, yeast and fermentation mashes.

This process has been studied in detail with a standardized clay suspension (5) and since then the same behavior has been shown to apply to pure cultures of bacteria, algae, and even to oil-in-water emulsions (3,9,12). A typical filtration plot for a clay suspension under constant pressure with different voltages applied is shown in Figure 6. With no voltage applied, i.e. under conventional filtration, rapid decrease in filtration rate is observed as a result of filter cake buildup. The application of 1.7 v/cm is already sufficient to show significant effects, i.e. to decrease the rate of filter cake formation. For any given flow rate there is a critical voltage above which constant flow rate is obtained as filter cake buildup is completely avoided. At this point the rate of electrophoretic removal of particles is equal to the linear rate of liquid flow through the filter, but opposite in direction. In first approximation, this critical voltage is given by equation (5) derived for filterable colloids. Though the expression is the same in both applications, it bears remembering that below critical voltage the effects are quite different. Filter cake buildup begins with particulate matter, while with filterable colloids incomplete elimination from the top effluent is observed.

The data in Figure 6 show that the application of the electrical field

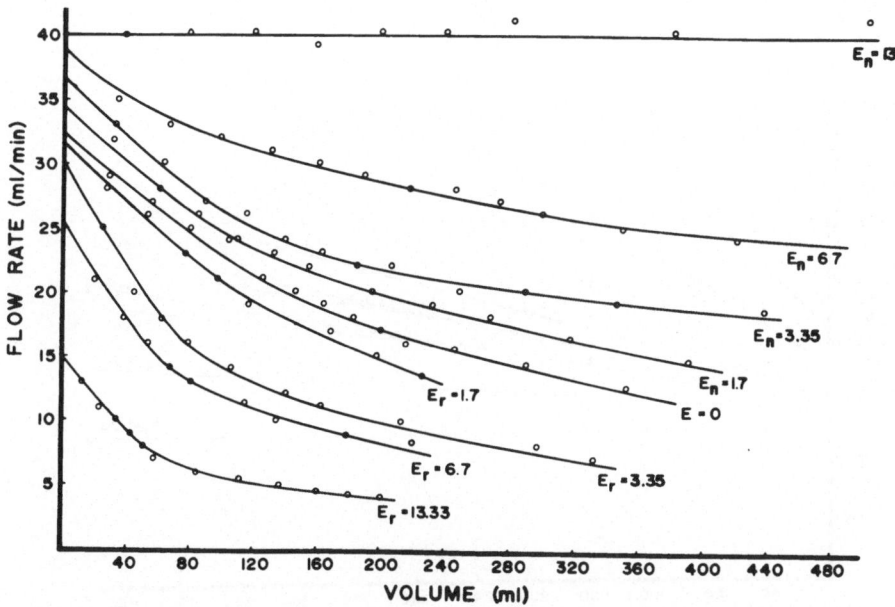

Figure 6: Filtration rate versus cumulative filtrate volume plots for bentonite suspensions (300 ppm clay in 450 ppm NaCl solution). 8" pressure differential, 300 cm^2 filter area. E_n - normal polarity voltage gradient; E_r - reverse polarity voltage gradient.

results not only in prevention of filter cake deposition, but also in increased initial filtration rates. These increased rates are observed also in absence of any particulate matter, and is due to the electroosmotic effect of the filter itself. It should be emphasized that the application of an electric field of reversed polarity, i.e. forcing the migration of the negatively charged particles towards the filter, has just the opposite effect, namely a more rapid clogging of the filter, and decrease in initial flow rates. This is shown in the bottom four lines of Figure 6.

Conventional filtration data at constant pressure are often expressed in terms of the D'Arcy equation, as modified by Sperry (19) or Carman (20)

$$T/V = \alpha \, \eta \, \nu \, V/2PA^2 + \eta R_m/PA \qquad (8)$$

which relates cumulative times T and filtrate volumes V to suspension viscosity η pressure P, and filter area A. R_m is the resistance to flow of the filter itself, α the resistance of the filter cake, and ν the volume of cake produced by unit volume of filtrate. According to this equation, a plot of T/V versus V should yield a straight line, where the intercept at time zero is a function of filter medium resistance, and the slope of the line a measure of filter cake buildup.

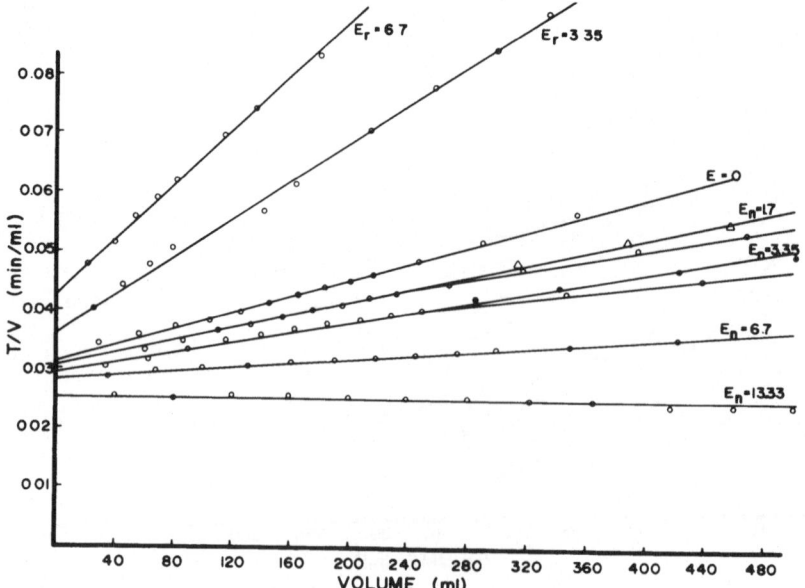

Figure 7: Carman equation plots of data contained in Figure 6. Circles - original results; triangles - data corrected for electro-osmotic effect of filter cake.

If the data of Figure 6 are replotted in terms of T/V versus V, the zero voltage plot gives a straight line in conformity with equation (8) as shown in Figure 7. This was the case with all suspensions so far studied in our experimental arrangement. With increasing voltage of the proper polarity, the straight lines are displaced, showing a lowering of intercept and decrease in slope, until at or above critical voltage horizontal lines are obtained, showing absence of filter cake formation. With reverse polarity, causing migration of clay towards the filters, the opposite effects are obtained. Moreover, lines showing some filter cake formation present a sharp break in slope, after certain amount of deposit is formed. The first effect, lowering of intercept, is due to decreased filter resistance, as a result of an electroosmotic effect of the filter itself. It is noticeable even in absence of particulate matter, i.e. with clear liquids. The second effect, decreased slope, is due to partial or complete removal of clay by electrophoresis. The break in the plots, is due to an additional electroosmotic effect of the filter cake, rather than filter, but it becomes noticeable only after a minimum depth of cake is deposited. It is significant that all three effects are additive and contribute to the efficiency of the electrofiltration process.

These effects necessitate corrections of above equation (8), and, in first approximation, these are:

a. In absence of particulate matter, electrofiltration rate Q_f is higher than the filtration rate $Q_{f,o}$ (obtained in absence of applied voltage) because of the electroosmotic flow Q_e caused by the zeta potential ζ of the filter itself. The well known Smoluchowski equation gives

$$Q_e = DI\zeta/4\pi\eta k = KI\zeta/k \tag{9}$$

where D represents the dielectric constant and K a constant for a given medium. Therefore, total flow is

$$Q_f = Q_{f,o} + Q_e = Q_{f,o} + KI\zeta/k \tag{10}$$

and the specific filter resistance R_m in electrofiltration becomes

$$R_m = R_{m,o}Q_{f,o}/Q_f = \frac{R_{m,o}}{1 + KI\zeta/kQ_{f,o}} \tag{11}$$

The substantial validity of this correction has been confirmed by calculations from experimental data of the factor $K\zeta = Q_e k/I$ which remained constant in a given system under a variety of conditions (5).

b. In the case of normal filtration all solids contained within the filtered volume are assumed to form the filter cake, but in electrofiltration the mass of the cake is decreased by the ratio

$$(V - \mu \, EAT)/V \tag{12}$$

and the slope of the Carman plots, i.e. the first term of equation (8), has to be multiplied by this factor. μEAT is the volume cleared by electrophoresis alone. By definition of critical voltage, above ratio can be also expressed as

$$(E_{cr} - E)/E_{cr} \, . \tag{13}$$

Combining these two corrections equation (8) becomes in case of electrofiltration:

$$\frac{T}{V} = \frac{\alpha \, \eta \, \nu \, V(E_{cr} - E)}{2PA^2 \, E_{cr}} + \frac{\eta R_{m,o} Q_{f,o}}{PA(Q_{f,o} + Q_e)} =$$

$$= \frac{\alpha \eta \nu \, (V - \mu \, EAT)}{2PA^2} + \frac{\eta \, R_{m,o}}{PA(1 + KI\zeta/kQ_{f,o})} \, . \tag{14}$$

Above critical voltage this equation is reduced to

$$\frac{T}{V} = \frac{\eta R_{m,o}}{PA(1 + KI\zeta/kQ_{f,o})} \tag{15}$$

and constant rate of filtration is obtained. This is the most efficient condition for electrofiltration.

c. The break in the plots of Figure 7 is caused by electroosmosis within the filter cake itself. The effect becomes noticeable only when a finite thickness of cake is deposited, but remains constant thereafter, as electroosmosis is independent of the thickness of the porous plug. Denoting as T_{cr} the time at which sufficient cake has been deposited to first give rise to this flow, and denoting as Q'_e the electroosmotic flow through the cake, we can obtain a corrected total volume of filtrate V', by subtracting from the actual volume the volume contributed by filter cake electroosmosis:

$$V' = V - (T - T_{cr})Q'_e \tag{16}$$

or

$$\frac{T}{V'} = \frac{T}{V - (T - T_{cr})Q'_e} \tag{17}$$

The factor Q'_e has been determined experimentally (5) by measuring flow rates of clear solvent with and without applied voltage, through preformed filter cakes. When the correction was applied to experimental data, the break in the plots vanished, as evident from the corrected curves in Figure 7.

D - ELECTROFILTRATION WITH CROSS-CURRENT FLUID EXCHANGE

A special variant of electrofiltration is illustrated schematically in Figure 8. A continuous stream of a suspension of particulate matter is flown between two filters of sufficiently fine porosity to mechanically contain the particles. A cross-current flow of a rinsing fluid is maintained across the compartment formed by the two filters. The fluid

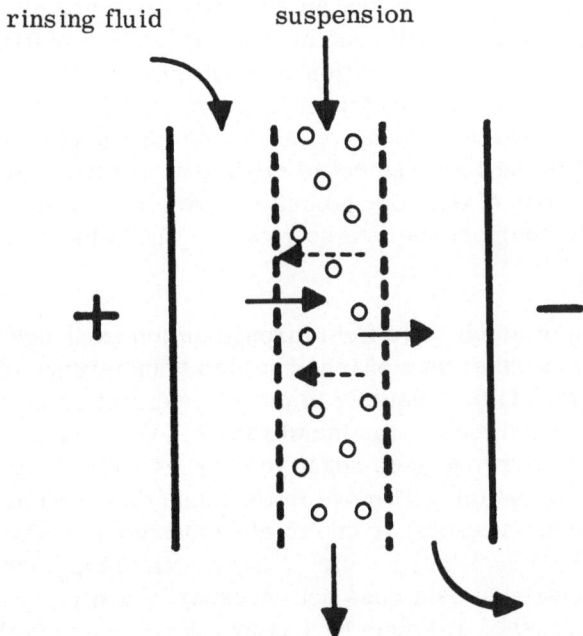

Figure 8: Cross-current rinsing of a suspension (red blood cells, for example) using electrofiltration. Suspension is contained between two filters. Solid arrows - direction of liquid flow; broken arrows - direction of electrophoretic migration.

penetrates the first filter by pressure alone, but clogging of the second filter is prevented by the application of the electrical field. Thus, electrofiltration occurs only on this second filter. Because of the cross-current nature of the fluid flow, efficient exchange of fluid is obtained. This process has been explored for the rinsing of erythrocytes. A current method of their preservation is through freezing in glycerine, which has to be washed out prior to their use. Millipore filters of 0.22μ porosity were employed. While normally red blood cells clog such filters immediately, continuous flow is obtained with electrofiltration. The rate of flow is completely conductivity dependent. If the necessary isotonicity is maintained with sodium chloride (0.9%), a cross-current flow rate of only 360 ml/hr per sq ft filter area is obtainable, but in isotonic glucose (5%), the rate of flow is in excess of 12,000 ml/hr per sq ft area.

E - ELECTROOSMOTIC CONCENTRATION AND/OR DESALTING

It is well known that efficient desalting can be achieved through electrodialysis, using an alternate series of anion and cation permselective membranes. To obtain this effect, it is not necessary to have both type of membranes present, but an alternate sequence of anion or cation selective membranes with neutral membranes will suffice. Mintz (21) for example, has studied the cation-neutral membrane transport depletion process, and has shown that it has some potential advantages over the usual bipolar arrangement. The reason for the ability of such a system to desalt is the requirement of electrical neutrality in solutions: it is sufficient that ions of only one polarity be retained by some of the membranes that its counterions be also retained within the same compartment.

The application of an electric field through an ion exchange membrane will not only cause transference of ions but also transference of the solvent. This solvent flux is usually in the same direction as that of the counterions (to the fixed ions of the membranes). For instance, in an anionic membrane, carrying fixed negative charges and selectively permeable to cations, water will move in the same direction as the cations. This solvent transport is due to electroosmosis within the membrane, conceived as a 'porous plug', but the cited Smoluchowski equation (9) for electroosmosis does not necessarily apply. The reasons for it have been discussed at length by Lakshminarayanaiah (22). No satisfactory theory exists and the treatment is largely empirical. The transport is usually expressed as moles of water transferred per faraday, $\bar{t}_w$, and has been studied with a variety of membranes and experimental conditions. It has been found that in general $\bar{t}_w$ decreases as the external concentration of counterions is increased. For example, typical data for a PSA membrane show that $\bar{t}_w$ decreases from 51.1 moles/faraday at 0.01N, to 37.1 at 0.1N, and to 11.0 at 1.0N external sodium chloride concentration. Furthermore, $\bar{t}_w$ decreases with decreasing water content of the membrane, or increase in membrane cross-linkage. The nature of the counterions is also of importance (22).

It has been discovered only recently (6, 8, 18) that these effects of ion exchange membranes, namely desalting and water transport, can be mimicked by using electrically neutral membranes, such as Visking membranes mentioned above, and a polyelectrolyte containing solution. Under the influence of the electrical field, the polyelectrolytes migrate towards the membranes, and the resulting polarization of charged groups against the neutral membranes will convey to it all the characters of an ion exchange membrane.

The arrangement employed is illustrated schematically in Figure 9, and is identical to that employed in conventional electrodialysis, only that electrically neutral Visking cellulosic membranes are employed. The double arrow indicates the direction of the flux of both, cations and water, across the Visking membrane into the polyelectrolyte compartment. The most surprising aspect is the rapidity of the water transfer, which is

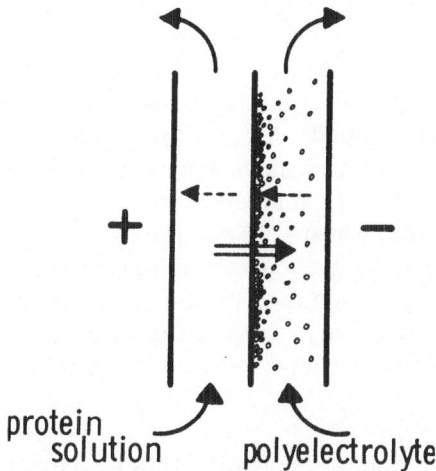

Figure 9: Schematic presentation of electroosmotic concentration and desalting. Broken arrow - direction of electrophoretic migration of polyelectrolyte and proteins; double arrow - direction of water flux and counterion transport. Concentration polarization of polyelectrolyte against the cathode facing membranes is indicated.

much higher than can be obtained by pressure or osmotic gradients alone, using these membranes. A number of polyelectrolytes have been studied, both soluble ones, and finely pulverized suspended ion exchange resins. Most effective were found to be polyacrylic acid (Rohm & Haas Co.), poly (ethylene-maleic acid) obtained through hydrolysis of its anhydride, commercially available under trade name EMA (Monsanto), and poly (methyl vinyl ether-maleic acid) also obtained from its anhydride, GANTREZ-AN (GAF). Their range of effective concentration seems to extend from 3-5% to below 0.1%, but their bulk concentration does not give true indication of their effective concentration, as they are electrophoretically retained in the cell (18).

Most work on this technique has been carried out for the purpose of concentration and/or desalting of proteins, enzymes and other biological materials (23). While proteins also cause an electroosmotic effect, it is small in comparison to that produced by the polyelectrolytes, and even very concentrated protein solutions can be handled without loss in efficiency. For instance, serum proteins can be readily concentrated to 20% solid content, and enzymes were concentrated or desalted without loss in biological activity. There is no membrane fouling or protein deposition on the dehydrating side of the membrane, as the proteins themselves have sufficient charge to migrate electrophoretically away from this membrane, i.e. they migrate in the same direction as the polyelectrolytes. The main advantage of this technique, when applied to proteins, is its large capacity and absence of membrane fouling.

Desalting and concentration occur simultaneously and are superimposed. However, the method is most efficient for desalting at low salt levels, while for dehydration, it is most efficient at high salt levels. If desalting without concentration is desired, constant volume may be maintained by addition of distilled water. If maximum rate of concentration is desired, concentrated salt may be added to maintain optimum salt levels. Optimum salt level is in the range of 0.05 to 0.1 molar NaCl. With a current density of 0.02 amps/cm^2, $\bar{t}_w$ is of the order of 50 moles/faraday, which is higher than that obtained at comparable concentration with formed ion exchange membranes.

No attempts have been made to optimize this process to desalting of brakish waters. Of possible advantage is that the cost of membrane forming is avoided, minimal membrane thickness may be achieved, and there is zero cross-linking and occluded volumes. The greatest obstacle is the progressive dilution of the polyelectrolyte, and a simple means of its recovery would be necessary.

F - ELECTRODEPOSITION ON MEMBRANES

A final membrane effect to be discussed is the electrodeposition of colloids. Coating of various materials through electrophoretic deposition is a widely practiced industrial process. In EMP, electrodeposition has usually a nuisance value, as it results in membrane fouling and shortens the useful life of a membrane assembly. It is fortunate that in most EMP processes, electrodeposition is avoided. This is largely due to the electro-osmotic effect described in the previous section. This electroosmotic water flux causes a sufficient backwash of membranes, preventing the deposition of most colloids and membranes emerge clean of deposits even after prolonged use. This is not the case with some of the systems studies, for instance casein in milk or some synthetic latices, which have insufficiently high charge density to cause strong electroosmosis through the membrane, and precipitate as a result of their low colloid stability (also associated with low zeta potential). In electrodialysis of whey, this can be partially prevented by turbulent flow.

In certain instances, however, advantage can be taken of the electro-deposition. Viruses in water are at too low concentration to be directly detectable through the usual tissue culture techniques. Concentration procedures are therefore necessary. Using bacteriophages as models of viral water pollution, it has been shown that they can be reversibly absorbed on cellulosic membranes through FFE, their concentration being too low to cause any electroosmotic effect or electrodecantation. They were detected by overlying the membranes, carrying the adsorbed bacteriophages, directly on a suitable growth medium. Alternately, they could be brought into solution by simple reversal of current polarity. The use of this technique has been proposed for virus detection in polluted waters (7).

APPLICATIONS TO BLOOD AND ITS COMPONENTS

The most important applications of EMP are in the realm of biological processing, particularly in their applications to blood and its various derivatives. A number of applications have been studied. Simplest is the fractionation of serum or plasma proteins by FFE, which can readily yield both γ- globulin and serum albumin (2-4, 25). γ- globulin is obtained by virtue of the fact that it is the slowest component of plasma, near isoelectric over the pH range 7-8. By slight variation in the conditions of the fractionation, such as pH, voltage, flow rate, and conductivity, the purity of the fraction can be varied at will, yielding either pure IgG, or mixtures of immune proteins, IgG, IgM, and IgA. Such a

broad fraction has been found to be an excellent source of preparation of
large quantities of IgM, by subsequent column chromatography (24). Serum
albumin is the fastest protein in plasma, and can be isolated either by conventional FFE, through depletion of all globulins, or through the use of
ultrafiltration membranes, as described above, section B. Intermediary
components can be obtained only in cascade operations.

The rate of processing achievable is largely dependent of the salt
content. With serum or plasma, flow rates are of the order of 350 ml/hr
per sq ft membrane area, but with plasma deionized through dialysis or
other procedures, flow rates of 1500 ml/hr have been achieved without
loss of resolution. The antibody content of the isolated γ - globulin is
not impaired, and antilymphocytic globulin isolated from plasma of
suitably immunized horses has been used clinically, for the prevention of
organ rejection following kidney transplantation (25).

Fractionation of plasma proteins is also possible with an <u>in vivo</u>

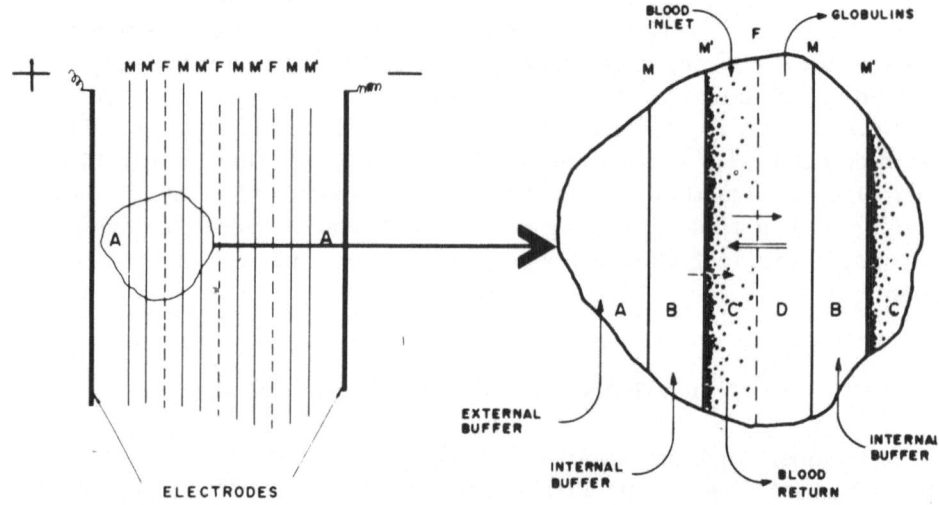

Figure 10: Schematic presentation of cell arrangement employed
in selective plasmapheresis. M, M' - membranes, F - filters,
A - external buffer compartment, B - internal buffer compartments, C - blood compartments, D - globulin output compartments. Solid arrows mark the direction of liquid flow, double arrow
shows the direction of electrophoretic migration, broken arrow
shows the direction of the electroosmotic water flux (4).

technique referred to as selective plasmapheresis (4). Heparinized whole blood is continuously circulated through an FFE apparatus in a manner similar to that used in artificial kidneys. Not only are plasma proteins fractionated, but also all corpuscular elements of blood are separated from the globulin fraction. It is the most demanding fractionation process, as it has to be carried out under physiologically acceptable conditions of pH and electrolyte balance, and without damaging the fragile corpuscular elements of blood. A membrane assembly as depicted in Figure 1 can be used, but the arrangement illustrated in Figure 10 is preferred. In this arrangement, each cell is separated from the next one by a channel formed by two dialyzing membranes, through which so-called internal buffer is circulated. These additional compartments permit better control of operating conditions, temperature, pH, conductivity, etc. The volume of the globulin fraction withdrawn from an animal in a single treatment lasting about 4 hours, may be in excess of the total blood volume of the animal at the beginning of the experiment. A complete dehydration of the blood would result. By fortuitous circumstance, the erythrocytes in blood exert a very strong electroosmotic gradient (see above section E), and the resulting water flux from the internal buffer into blood is of the same order as the rate of globulin separation. Automatic regulation of blood volume is therefore obtained. In selective plasmapheresis, electrophoretic fractionation of plasma proteins, electrofiltration of the corpuscular elements, and electroosmotic flux of water are all super-imposed.

The overall flow pattern is illustrated in Figure 11. In a single treatment, radical depletion of circulating globulins can be obtained, thereby substantially modifying the composition of the animal's blood. This is illustrated in Figure 12, where the composition of plasma of a sheep is reported, from which 40 and 70% of total circulating globulins were withdrawn on two consecutive days. Selective plasmapheresis has been applied to dogs, sheep and horses (2-4, 25-28). It is of importance because immune antibodies are contained within the γ- globulin fraction. It has been shown that it is possible to modify the immune reponse of an animal through this procedure of antibody depletion, using hetero-graft kidney transplantation as model (4, 27, 28).

In above schematic arrangement of membranes and filters, electro-dialysis is by necessity superimposed upon the fractionation process. It is possible to employ a similar arrangement for blood electrodialysis alone, by omitting the filter and the globulin compartment D, of above Figure 10. Such an arrangement has been studied extensively as an electrodialytic artificial kidney. Usual artificial kidneys operate on the principle of passive dialysis, while the application of an electric field

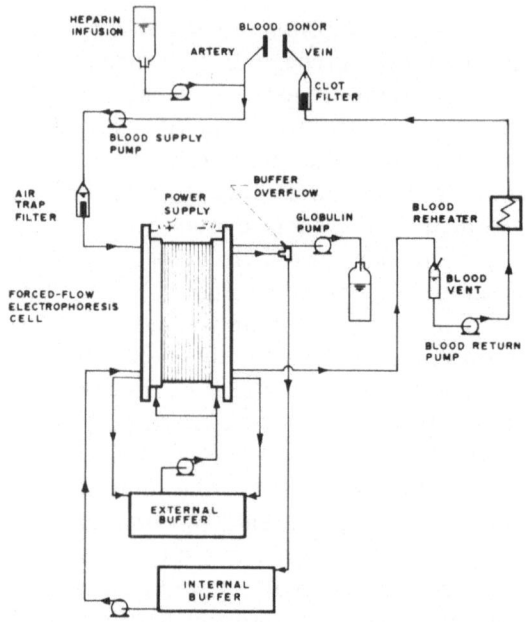

Figure 11: Overall presentation of flow patterns in selective plasmapheresis - the electrophoretic fractionation of whole blood maintained in extra-corporeal circulation (4).

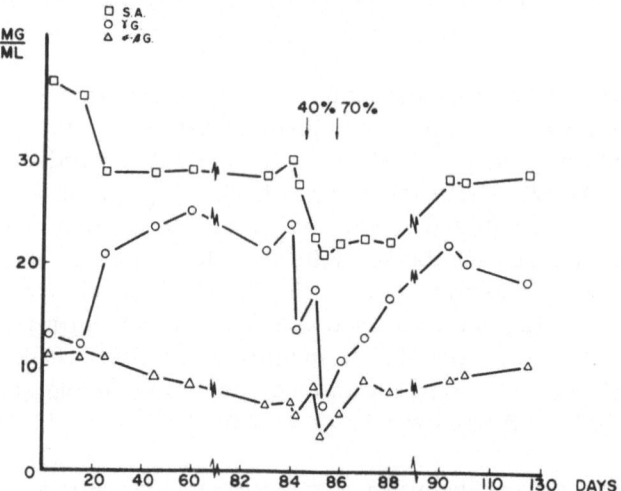

Figure 12: Effect of selective plasmapheresis on plasma composition of a sheep. The sheep was hyperimmunized during the first 60 days, and subjected to plasmapheresis on two consecutive days, with a withdrawal of 40 and 70% of circulating γ-globulins. Squares - serum albumin; circles - γ - globulin; triangles - α - and β - globulins.

imposes an additional parameter. Fivefold acceleration of removal of ionized species from blood has been obtained (6).

The last EMP applied to blood derivatives was the washing of suspensions of erythrocytes, discussed in above section D.

APPLICATIONS OF EMP TO WATER AND WASTE PURIFICATION

The feasibility of water and waste treatment through EMP has been established in several recent studies comprising tertiary treatment of sewage effluents, river, well, and flood waters, some industrial wastes, and artificial mixtures containing representative pollutants (3,7-10,12-14). Three objectives can be achieved simultaneously - complete removal of suspended materials, such as algae and clay, through electrofiltration - effective elimination of soluble colloidal pollutants, such as humic acid, proteins, colloidal iron, etc., through FFE - and also substantial desalting, as a result of electroosmotic effects. The main purpose is removal of colloidal impurities, and with increasing industrial discharge and reuse of water, the range of pollutants can be very broad. In a number of studies, Black and his co-workers (29,30), Riddick (31,32) and others (33,34) have shown that most colloids in water and sewage carry large negative charges. Conventional water treatment through coagulation necessitates reduction of these charges, in accordance with the general principles of the dependence of colloid stability on the zeta potential (14,29-32). In EMP, direct advantage is taken of these large charges for removal of colloids (14). As the conductivity of many waters is low, this is not only technically feasible but also economically possible.

As an indication of results that can be expected, Table I and II summarize some data obtained on samples supplied, processed and analyzed through the courtesy of Holyoke Water Power Co., in Holyoke, Mass (12). Significant reduction in total and dissolved solids has been achieved. Because of the desalting in the first step, reprocessing in a second or third pass becomes most economical, and yields water of excellent quality.

The application of EMP to water and waste purification requires largest volume capabilities at lowest possible cost, and this point merits closer evaluation.

Equation 4, on previous pages, purports to show the relation between purification and voltage or current applied. In the present context, more important is to express the performance in terms of total power $W = I^2R$,

Table I

TREATMENT OF HOLYOKE WATER SAMPLES (12)[1]

	Connecticut River Water			Raw Water, Well #1			
	raw water	treated		raw water	treated		
		1x	2x		1x	2x	3x
	(ppm)	(% reduction)		(ppm)	(% reduction)		
Hardness, Total	34	62	100	246	71	96	100
Alkalinity, Total	29	45	76	45	38	82	89
Turbidity	14	86	100	35	97	100	100
Chlorides	7.1	58	85	13.9	68	95	97
Iron, Total	0.44	70	100	3.20	96	98	99
Manganese	0.00	-	-	1.84	73	100	100
Color	45	89	100	100	100	100	100
Suspended Solids	34	94	94	11	45	45	100
Volatile Suspended Solids	13	85	85	11	45	45	100
Fixed Suspended Solids	21	100	100	0	-	-	-
Dissolved Solids	83	67	93	467	71	94	96
Volatile Dissolved Solids	22	60	91	124	73	96	96
Fixed Dissolved Solids	61	69	93	343	70	93	97
Phosphates	0.96	46	73	0.26	50	70	81

------
[1] All waters processed and analyzed by courtesy of Holyoke Water Power Co., Holyoke, Mass.

Table II

TREATMENT OF HOLYOKE WASTE SAMPLES (12)[1]

| | Paper Plant Effluent | | Holyoke Sewage Treatment Plant Effluent | |
	raw water (ppm)	treated (% reduction)	raw water (ppm)	treated (% reduction)
Biochemical Oxygen Demand	210	99	139	40
Alkalinity, Total	21	86	94	77
Suspended Solids	686	99	46	100
Volatile Suspended Solids	536	99	46	100
Fixed Suspended Solids	150	100	0	-
Dissolved Solids	190	43	146	62
Volatile Dissolved Solids	170	41	96	76
Fixed Dissolved Solids	20	55	50	36

------
[1] All waters processed and analyzed courtesy of Holyoke Water Power Co., Holyoke, Mass.

where R, the total resistance of a multicompartment cell pack, is

$$R = (Nd/kA)f \qquad (18)$$

where d is the thickness of a single cell and f the polarization factor, i.e. increase in cell pack resistance over that calculated on the basis of water conductivity, due to polarization of cell content. Total power, W, accordingly, is

$$W = \frac{Q_f^2 F^2 \; kdf}{\mu^2 \; NA} \qquad (19)$$

whereby, as previously, all factors are expressed in cgs units. One may wish to convert these to the usual engineering units, and obtains the power consumption per 1,000 gal as being:

$$kwhr/1,000 \; gal = \frac{Q_f F^2 kdf}{\mu^2 NA} \times 3 \times 10^{-3} \qquad (20)$$

The factor 3×10^{-3} results from the conversion of units, and the dimensions employed are listed below. Based on above equation, calculated power requirements are presented in Table III, as a function of specific water conductivity, and various production rates per unit membrane area. The following values were assumed:

Colloid mobility of $\mu = 15 \times 10^{-5}$ cm²/volt sec., corresponding to the lower range of values observed in many waters (29-34).

Cell thickness d = 0.14". This is about twice the thickness customary in electrodialysis, but is necessary if particulate matter is to be processed.

Purification factor F = 0.95. Up to this limit, equation 4 is usually applicable.

Polarization factor f = 2. This is a rather high value, and presumes high current density.

The assumed production rates of 1 and 2 gal/hr per sq ft filter area cover the range of flows encountered in pilot runs. At these flux rates, 400 or, respectively, 200 sq ft filter area would be sufficient to produce 10,000 gal per day.

Table III

ESTIMATED POWER REQUIREMENTS FOR WATER
PURIFICATION BY EMP

Flow Rate/ Membrane Area	Power Requirement for Waters of Conductivity (kwhr/1,000 gal.)		
	100 μ mho/cm	300 μ mho/cm	1000 μ mho/cm
1 gal/hr sq ft	3.5	10.5	35
2 gal/hr sq ft	7	21	70

Of all above factors, greatest variability is exhibited by the conductivity of the water being processed, and this will determine the applicability of EMP. The composition of waters from different sources is notoriously variable. Nevertheless, there seem to be relatively fixed empirical ratios between specific conductivity and water compositions (35):

total dissolved solids (ppm)/micromhos per cm = 0.64 and
micromhos per cm/total cation concentration (meq/l) = 100.

To estimate the applicability of EMP to various surface waters, data on composition of a number of rivers, lakes, and spring waters were grouped according to total concentration of main cations and are tabulated in Table IV in the expectation that the original list (36) was a rather representative cross section of available water sources. It would appear that about 50% of the available sources are amenable to water treatment by EMP, having an estimated total cation concentration of less than 3 meq/l, or conductivities of less than 300 micromhos/cm.

The closest analog to EMP is electrodialysis, and a comparison might be of interest. Pertinent data on five actual installations for water desalting, taken from Spiegler (37) are listed in Table V. Obviously, electrodialysis is applicable to far saltier solutions than EMP, which is primarily designed for colloid removal and the two methods should not be viewed as competitive but as complementary. Equipment and general costs may be comparable in the two processes, though EMP uses considerably cheaper membranes and filters. It should be emphasized, however, that above data for EMP are purely hypothetical extrapolations of small scale laboratory work, and no large installations were tried. The longevity of membranes and filters in an assembly is the primary unknown.

Table IV

WATER SOURCES GROUPED ACCORDING TO TOTAL CATIONIC CONCENTRATION
(From Nordell (36))

Sources	Number of Samples Having Concentrations In Ranges					
	0-1 meq/l	1-2 meq/l	2-3 meq/l	3-4 meq/l	4-5 meq/l	over 5 meq/l
100 River Samples	19	25	9	9	11	27
50 Lake Samples	21	7	11	4	0	7
44 Spring Samples	8	7	5	7	4	13

Table V

POWER REQUIREMENT FOR SOME ELECTRODIALYSIS PLANTS (FROM SPIEGLER (37))

Location	Union of South Africa	Tobruk, Lybia	Salt Lake City, Utah	Coalinga, California	Neot Hakikar, Israel
Designer	S.A.C.S.I.R.	Wm Boby and Co.	Ionics, Inc.	Ionics, Inc.	Negev Rsch. Inst.
meq/l (NaCl)	48	79	56	41	51
gal/hr sq ft	0.35	0.31	0.68	0.40	0.38
kwhr/1000 gal	7.75	13.3	13.6	6.3	17.0

SUMMARY

Electrophoresis, electrodialysis and electroosmosis, as well as new processes which could be termed electrofiltration and electroadsorption, can all be studied utilizing relatively simple arrangements of membranes and filters, with a d.c. electrical field imposed across them. Because of the overlap of effects it is impossible to draw sharp distinctions, and it is best to group the above interrelated operations under the term of electrokinetic membrane processes. Their primary usefulness is probably in the treatment of complex materials of biological origin, for purposes of fractionation, concentration, and/or desalting. A system of large scale human plasma fractionation can yield the desired end products, mainly immunoglobulins and albumin, by above processes alone, or in combination with the conventional alcohol fractionation techniques. The processes are sufficiently efficient, however, that they could be applied to purely industrial uses, such as water or waste purification. The main factor affecting cost efficiency is the salinity of the feed, and fluxes of up to 3 gal/hr sq ft membrane area can be obtained. Of unique interest is the continuous non-clogging electrofiltration of particulate suspensions, such as clay, algae or emulsions, the electrical field preventing the formation of filter cake. Thus, in electrokinetic processes, electrical driving forces replace pressure gradients: the additional parameter of electrical charge of solutes or membranes adds an element of flexibility not present in reverse osmosis or ultrafiltration.

REFERENCES

1. Rickles, R. N.: 'Membrane Technology and Economics', Noyes Development Corp., Park Ridge, N. J., 1967.

2. Bier, M.: 'Preparative Electrophoresis without Supporting Media', in 'Electrophoresis' (ed. M. Bier), Academic Press, New York, N. Y., 1959, p.263.

3. Bier, M.: 'Forced-flow Electrophoresis and its Biomedical Applications', in 'Membrane Processes for Industry', Southern Research Institute, Birmingham, Ala., 1966, p.218.

4. Bier, M., Beavers, C. D., Merriman, W. G., Merkel, F. K., Eiseman, B., and Starzl, T. Z., Trans. Amer. Soc. Artif. Int. Organs 16, 325, 1970.

5. Moulik, S. P., Cooper, F. C., and Bier, M., J. Colloid Interface
    Sci., 24, 427, 1967.

6. Bier, M., Bruckner, G. C., and Roy, H. E., Trans. Amer. Soc. Artif.
    Int. Organs, 13, 227, 1967.

7. Bier, M., Bruckner, G. C., Cooper, F. C., and Roy, H. E.: 'Con-
    centration of Bacteriophage by Electrophoresis', in 'Transmission of
    Viruses by the Water Route', (ed. G. Berg), Interscience Publishers,
    New York, N. Y., 1967, p. 57.

8. Bier, M.: 'Electrophoretic Membrane Processes', in 'Symposium on
    Electrodialysis', Electrochemical Society, Boston, Mass., 1968.

9. Cooper, F. C.: 'Water Purification by Forced-Flow Electrophoresis',
    Thesis, University of Arizona, 1964.

10. Hannig, K.: 'Preparative Electrophoresis', in 'Electrophoresis', Vol.
    II, (ed. M. Bier), Academic Press, New York, 1967, p.423.

11. Tiselius, A., and Svensson, H., Trans. Faraday Soc. 36, 16, 1940.

12. Bier, M., and Moulik, S. P.: 'Water Purification by Large Scale
    Electrophoresis', Proceedings 3rd Amer. Water Resources Conference,
    Urbana, Ill., 1967, p.524.

13. Bier, M.: 'Water Treatment by Electrophoresis' in Proceed. 13th
    Annual Conf. Water for Texas, Texas A&M University, College
    Station, Texas, 1968.

14. Bier, M., and Cooper, F. C.: 'Electrical Phenomena at Surfaces',
    in 'Principles and Applications of Water Chemistry', (eds. S. D.
    Faust and J. V. Hunter), Wiley, New York, 1967, p.217.

15. Pauli, W., Biochem. Z. 152, 355, 1924.

16. Cann, J. R., Kirkwood, J. G., Brown, R. A., and Plescia, O. J., J.
    Am. Chem. Soc. 71, 1603, 1949

17. Polson, A., Biochem et Biophys. Acta 11, 315, 1953.

18. Bier, M., U. S. Patent Application.

19. Sperry, D. R., Chem. Metal Eng. 15, 198, 1916; ibid 17, 161, 1917.

20. Carman, P. C., Trans. Inst. Chem. Engrs. (London) 16, 168, 1938.

21. Mintz, M. S.: 'The Cation-Neutral Membrane Transport Depletion
    Process', in 'Symposium on Electrodialysis', Electrochemical
    Society, Boston, Mass., 1968.

22. Lakshminarayanaiah, N.: 'Transport Phenomena in Membranes',
    Academic Press, New York, N. Y., 1969, p.242.

23. Extensive work on this process has been carried out at Canalco, Inc.
    Rockville, Md., and at Philips Roxane, Inc., St. Joseph, Mo.

24. Cozine, Wm. S., private communication.

25. Moberg, A. W., Gewurz, H., Simmons, R. L., Gunnarsson, A.,
    Merkel, F., and Najarian, J. S., Surgical Forum 20, 261, 1969.

26. Watt, J. G., Mackie, W. S., Fell, B. F., Logan, E. F., and
    Mitchell, B., Res. Veterinary Sci. 11, 168, 1970; Logan, E. F.,
    Stenhouse, A., Watt, J. G., and Clark, A. E., ibid., in press.

27. Merkel, F. K., Bier, M., Beavers, C. D., and Starzl, T. E.,
    3rd Intern. Congress Transpl. Soc., The Hague, Netherlands, 1970,
    p. 183 (abstract); Surgical Forum 21, 261, 1970.

28. Moberg, A. W., Shons, A. R., Gewurz, H. Mozes, M., and
    Najarian, J. S., 3rd Intern. Congress Transpl. Soc., The Hague,
    Netherlands, 1970, p.185 (abstract); Shons, A. R., Jetzer, T.,
    Moberg, A. W., and Najarian, J. S., Surgical Forum 21, 262, 1970.

29. Pilipovich, J. B., Black, A. P., Eidsness, F. A., and Stearns, J. W.,
    J. Am. Water Works Assoc. 50, 1467, 1958.

30. Black, A. P., Singley, J. E., Whittle, G. P., and Maulding, J. S.,
    ibid. 55, 1347, 1963.

31. Riddick, T. M., Chem. Eng. 68, 13, 121, 1961.

32. Riddick, T. M.: 'Control of Colloid Stability through Zeta Potential',
    Livingston Publishing Co., Wynnewood, Pa.,1968.

33. Faust, S. D. and Manger, M. C., Water Sewage Works 111, No. 2, 1964.

34. Jorden, R. M., J. Am. Water Works Assoc. 55, 771, 1963.

35. Camp, T. R.: 'Water and its Impurities', Reinhold, New York, N. Y., 1963, p.130.

36. Nordell, E.: 'Water Treatment', 2nd Ed., Reinhold, New York, N. Y., 1961.

37. Spiegler, K. S.: 'Salt Water Purification', John Wiley, New York, N. Y., 1962.

INDUSTRIAL ULTRAFILTRATION

Robert L. Goldsmith, Richard P. deFilippi, Sohrab Hossain,
and Robert S. Timmins
Abcor, Inc., 341 Vassar Street, Cambridge, Mass. 02139

INTRODUCTION

During the past decade reverse osmosis and ultrafiltration have been advanced from laboratory curiosities
to important industrial unit operations. Practical
applications have gone beyond the production of potable
water and include a broad spectrum of solution concentrations and/or fractionations.

Reverse osmosis and ultrafiltration are similar
processes since both use a semipermeable membrane as
the separating agent and pressure as the driving force
to achieve separation. There are important differences,
however, which lead to different applications, process
conditions and equipment for each of the two processes
as shown in Table I. This paper deals with ultrafiltration and its industrial applications.

In an ultrafiltration process (Figure 1) a feed
solution is introduced into a membrane unit, where solvent and certain solutes pass through the membrane under
an applied hydrostatic pressure. Solutes whose sizes
are greater than the pore size of the membrane are retained and concentrated. The pore structuré of the
membrane thus acts as a molecular filter, passing some
of the smaller size solutes and retaining the larger
size solutes. The pore structure of this molecular filter
is such that it does not become plugged because the solutes
are rejected at the surface and do not penetrate the membrane. Furthermore, there is no continuous build-up of

Table I

DIFFERENCES BETWEEN REVERSE OSMOSIS AND ULTRAFILTRATION

	Reverse Osmosis	Ultrafiltration
Size of solute retained	Molecular weights generally less than 500 to 1000	Molecular weights generally over 1000
Osmotic pressures of feed solutions	Important, can range to over 1000 psi	Generally negligible
Operating pressures	Greater than 100 psi, up to 2000 psi	10 to 100 psi
Nature of membrane retention	Diffusive transport barrier; possibly molecular screening	Molecular screening
Chemical nature of membrane	Important in affecting transport properties	Unimportant in affecting transport properties so long as proper pore size and pore size distribution are obtained

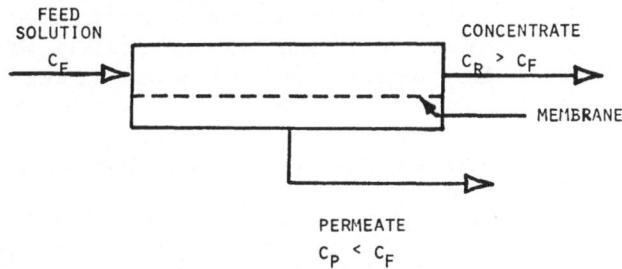

FIGURE 1: ULTRAFILTRATION FLOW SCHEMATIC

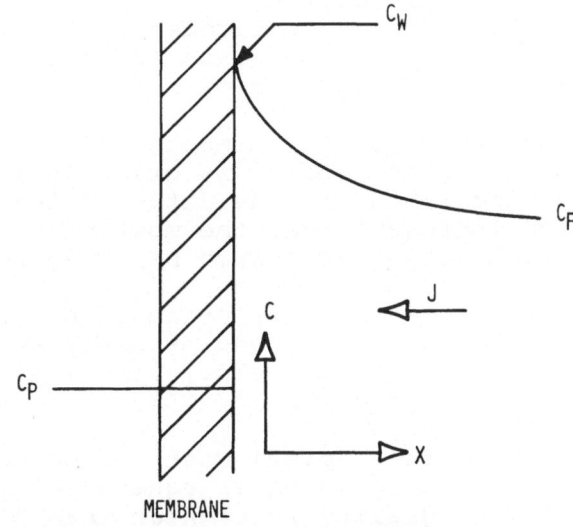

FIGURE 2: CONCENTRATION POLARIZATION SCHEMATIC

a filter cake which has to be removed periodically to re-
store flux through the membrane.   Many ultrafiltration
applications involve the retention of relatively high
molecular weight solutes, accompanied by the removal
through the membrane of lower molecular weight impurities
so that both concentration and purification are achieved.

Broadly speaking, the applications for ultrafiltration include those where it is necessary to concentrate,
fractionate or purify a molecular or colloidal solution.
In almost every case there is an alternative method of
performing these operations so that the membrane markets
will be those applications where the special properties
of the membrane process offer a functional or an economic
advantage or both.   The particular advantages offered
by the membrane process are:

1. The process is very gentle and requires no
    heat or chemical additions which can harm
    the properties of such materials as
    pharmaceuticals, foods and colloids.
2. The process is economical at both small
    and large sizes, because of its modular
    nature.
3. The process is very simple to operate,
    since it involves primarily the pumping
    of liquids.
4. The process is quite versatile at carrying
    out more than one function as in the case
    of simultaneous fractionation and purifi-
    cation.
5. Energy requirements of the process are
    quite low since operation proceeds through
    the pumping of liquids.

A comparison of the membrane ultrafiltration process
with competitive processes is given in Table II. A
review of the Table and the detailed features of ultrafiltration indicates that the most promising applications
for membrane processes should be those where either:

1. The product produced by the membrane has
    some improved property such as purity,
    activity, etc., versus an alternate tech-
    nique; or

2. The capital cost for competitive processes
    to produce an equivalent yield and product
    quality is greater than $1-2 per gallon
    per day capacity and the operating cost
    is greater than $1-2 per thousand gallons
    of water processed.

ULTRAFILTRATION THEORY

Concentration Polarization

In any ultrafiltration system, solution bulk flow
toward the membrane accompanied by solute rejection at the
membrane surface results in an elevated concentration of
rejected solutes at that point, known as concentration
polarization, illustrated in Figure 2. The solute concentration is shown as a function of position in the
vicinity of the membrane; C_F, C_W, and C_P are the solute
concentrations in the bulk feed solution, at the membrane
surface, and in the permeate solution. The ultrafiltration flux is represented by J. Solute transport toward
the membrane is balanced, at steady state, by solute

passage through the membrane and mass transport back
into the bulk feed solution. This model has been previously described (17,18).

In industrial ultrafiltration equipment the feed
solution is pumped past the membrane surface at a high
flow rate in order to reduce concentration polarization.
The greater the feed solution velocity, or Reynolds
number, the greater the shear at the membrane surface
and, as will be demonstrated, the greater the ultrafiltration rate. The optimum Reynolds number is determined
by a cost balance which involves both power costs and
membrane area costs.

According to Brian [6], concentration polarization
in turbulent flow ultrafiltration can be described in
terms of a mass transfer coefficient, k.

$$J = k \; \ln\left(\frac{C_W - C_P}{C_F - C_P}\right) \tag{1}$$

Relationships for k for different systems (laminar and
turbulent flow) have been described previously[8].
For high Schmidt number turbulent flow, the mass transfer
coefficient has been empirically correlated with various
system parameters [15]:

$$\frac{kd}{D} = 0.0096 \; Re^{0.91} Sc^{0.35} \tag{2}$$

Of importance are the system hydraulic diameter (d),
solute diffusivity (D), operating Reynolds Number (Re)
which is proportional to the flow rate through the membrane
system, and the Schmidt number (Sc). This relationship
will be discussed further in the light of data presented
for ultrafiltration in tubular membrane systems.

Flux Limitations

Osmotic Pressure Limitation. In ultrafiltration,
due to the low diffusivities of macrosolutes, concentration polarization is of major importance. For many
applications, high osmotic pressures at the membrane-
solution interface, resulting from concentration polarization, result in solution fluxes substantially lower
than the inherent membrane water flux.

For an ultrafiltration membrane, water flux

Table II

PROCESS COMPARISONS

Function	Competitive Process	Advantages of Membrane Process	Advantages of Competitive Process
Concentration	Evaporation	1. More gentle 2. No heat required 3. Cheaper for small plants	1. Can produce higher concentration 2. Cheaper for very large plants 3. Existing equipment often available
	Precipitation	1. More gentle 2. No chemical required 3. More effective for dilute solutions	1. Cheap for many products where functional properties not affected 2. Low capital investment required
Fractionation	Adsorption/Ion Exchange	1. Works over very wide range of molecular weights 2. No regenerants required	1. Often highly selective and can handle difficult separations
	Distillation	1. No heating required 2. Cheaper for small sizes 3. Wide range of molecular weights handled	1. Good selectivity 2. Economical in large sizes

Table II (Continued)

PROCESS COMPARISONS

Function	Competitive Process	Advantages of Membrane Process	Advantages of Competitive Process
Purification (Removal of pollutants from water)	Biological Oxidation	1. Removes both inorganic and organic pollutants 2. Recovers by-products 3. Cheaper for small-sized plants and high concentration	1. Very cheap for medium to large plants and moderate to low concentrations 2. Broadly used technique and many facilities available
	Adsorption	1. Works over very wide range of molecular weights 2. No regenerants required	1. Cheap for low concentration of pollutants 2. Can give very pure product water

is described by the relationship

$$J = A\Delta P \tag{3}$$

where A, the membrane constant, is dependent on temperature, and independent of pressure over the normal operating range. The hydrostatic pressure driving force is given by ΔP. When a macromolecular solution is ultrafiltered, flux is given by

$$J = A(\Delta P - \Delta \pi) \tag{4}$$

The osmotic pressure difference across the membrane, $\Delta \pi$, is the difference between the osmotic pressure at the membrane surface, π_W and the osmotic pressure of the permeate, π_P.

$$\Delta \pi = \pi_W - \pi_P \tag{5}$$

Based on molecular weight considerations, the osmotic pressures of macromolecular solutions would appear to be insignificant. However, for polymer solutions with concentrations over 1 wt.%, due to the importance of the second virial coefficient term, osmotic pressures exceeding 10 to 50 psi are not uncommon. For relatively dilute solutions (up to a few percent solute) the osmotic pressure can be related to concentration by Equation 6,

$$\pi = \frac{RTC}{M}\left(1 + \frac{\Gamma}{2}C\right)^2$$

This relationship [11] incorporates both the second virial coefficient, Γ, as well as an approximation for the third virial coefficient.

Experimental results for the ultrafiltration of solutions of polyethylene glycol and Dextran fractions which illustrate this type of flux limitation have been described elsewhere [13].

Gel Formation Limitation. In certain cases, the solute wall concentration, C_W in Figure 2, can reach a solubility limit, and solute precipitation onto the membrane surface can occur. This condition, described by Michaels [5,17,18], is promoted both by a low solubility limit of the macrosolute and factors which reduce the importance of the osmotic pressure effect. Osmotic pressure considerations assume less importance as the molecular weight of the solute increases (lower osmotic pressure per unit concentration), and with increasing operating pressure. The formation of a gel or "scale" on the membrane surface is sometimes only very slowly reversible, and mechanical and/or strong chemical cleaning is required for its removal.

For systems in which gel formation is reversible, that is, a rapid equilibrium is established between the solution and the gel layer at their interface, a dynamic equilibrium can exist. At steady state, the combined hydraulic resistance of the membrane and the gel layer is such that solute transport to the gel layer by bulk flow equals solute removal by passage through the membrane and mass transport back into the feed solution. For this case, flux is given by a relationship identical in form to Equation 1,

$$J = k \ln \frac{C_G - C_P}{C_F - C_P} \tag{7}$$

In this equation, the gel concentration, C_G, replaces C_W.

The differentiation between an osmotic pressure limitation and a gel formation limitation is sometimes difficult to make in practice. For both cases, flux is found to increase proportionally with k and to decrease in proportion to ln $(C_F - C_P)$. As has been shown for the osmotic pressure model [13], $(C_W - C_P)$ varies only slightly with changing feed concentration. The same is true for the gel formation case, in which C_G is fixed by the chemistry of the system. Some differences, however, have been observed: first, in gel forming systems, flux often declines slowly with time due to gel "compaction". This is generally not observed in non-gelling systems. Second, in non-gelling systems, a flux-versus-ln$(C_F - C_P)$ plot extrapolates to zero flux at a concentration below that of known gel formation. Solutions with concentrations above this level, with very low viscosities compared to gels, have shown zero flux [13]. This can be explained only by an osmotic pressure difference across the membrane exceeding the hydrostatic pressure driving force.

In both Equations 1 and 7, it is necessary to choose an appropriate value for the mass transfer coefficient, incorporating average values of the solution viscosity (μ) and the macromolecular diffusivity (D). Obviously both could vary substantially in an ultrafiltration system, passing from the bulk feed solution through the

polarization layer up to the gel layer. Research in our
laboratory and by others (7,12) is currently underway
to include these variations in mass transfer, and corresponding flux, considerations.

Equations 1 or 7 cannot presently be used to
accurately predict ultrafiltration fluxes. Nevertheless,
they can be used empirically to correlate data. A
"measured" rather than "calculated" value of k can be
used for actual design purposes. At present, plant
designs require pilot scale data, both for flux and membrane life.

Rejection

Evidence indicates that ultrafiltration membranes
are microporous, and membrane transport of both solute
and solvent is by viscous flow through the pores (1,4).
This is to be contrasted with "diffusive" membranes in
which transport is by molecular diffusion of individually
dissolved molecules. Ultrafiltration membranes of
commercial interest, i.e., so-called high flux membranes,
are prepared with an asymmetric pore structure. Pore
radii are smallest at the membrane surface which contacts
the solution being ultrafiltered (18). Passing into the
membrane from this surface, pore radii become progressively
larger. Consequently, a molecule entering a pore is
carried through the membrane with less and less hindrance.
This type of pore structure is quite different from that
of ordinary membrane filters, where particles enter the
pores, collect, and build up within the filter, due to
internal constrictions. This leads to pore plugging with
accompanying drastic reductions in filtration rates.

The rejection properties of ultrafiltration membranes
are related to steric factors inhibiting macromolecular
entry into the pores. For electrically charged macromolecules and/or membranes, electrostatic factors are
also important.

In practice, a membrane is selected experimentally,
such that it has the desired retention/fractionation properties. A general requirement is that close to 100% retention of the desired solute is obtained. When fractionation is a desired objective, 0% retention of another
solute is sought. In certain cases complete fractionation
(100%/0% retention) cannot be obtained due to a relatively
small difference in molecular sizes of the species to be
separated. For high fractionation efficiency molecular

sizes should differ by a factor of two or more.

In measuring rejection experimentally it is
important to use the exact mixture of interest rather
than measuring the rejection of the individual components in pure solutions due to the interactions which
occur between the various components. For example,
a gel layer of one higher molecular weight solute can
act as a dynamic rejecting membrane for lower molecular
weight materials (10,20,22). When this occurs, operation
at a lower pressure can result in greater fractionation
efficiency.

MEMBRANE SYSTEM CONFIGURATIONS

Membrane systems are commercially available or potentially available in several configurations. Most of
these have been developed for water desalination (reverse
osmosis), and some have been adapted to ultrafiltration.
While mechanical strength is an important consideration
for high-pressure reverse osmosis systems, it assumes
less importance for low-pressure ultrafiltration. Instead, mass transfer becomes exceptionally important and,
correspondingly, system design and geometry.

Listed in Table III are different membrane configurations, divided into turbulent flow and laminar flow
devices. Advantages and disadvantages of each for ultrafiltration applications are identified, and the status of
their commercial development is indicated.

ULTRAFILTRATION APPLICATIONS

Potential applications of ultrafiltration have been
under discussion in the literature for some time. Only
recently, however, have several of these become operating
realities. The purpose of this section is to describe
some of the more important current uses of ultrafiltration, and to indicate future areas where this process
will be useful on an economic, commercial basis.

Cheese Whey

About 22 billion pounds of cheese whey are
produced annually in the United States, of which some-
what less than half is treated to recover by-products
and/or to abate water pollution. The remainder is discharged by various means and constitutes a major source
of water pollution. The total whey production contains

Table III

MEMBRANE CONFIGURATIONS FOR ULTRAFILTRATION

System	Advantages	Disadvantages	Commercial Status
Turbulent Flow			
A. Tubular	1. easily cleaned for food applications; accepted for dairy and food processing 2. well-developed equipment 3. individual tubes can be replaced	1. high holding volume per unit membrane area 2. presently moderately expensive ($\sim$10-20\$/ft^2) 3. moderate parasitic pressure drop in tube connections (turnarounds)	commercially available
B. Flat Leaf*	1. low holding volume per unit membrane area 2. well-developed equipment	1. susceptible to plugging at flow stagnation points; possibly difficult to clean for food, dairy and pharmaceutical applications 2. presently expensive ($\sim$100\$/ft^2) 3. moderate parasitic pressure drop losses 4. entire membrane module must be replaced in case of failure	commercially available

Table III (Continued)

MEMBRANE CONFIGURATIONS FOR ULTRAFILTRATION

System	Advantages	Disadvantages	Commercial Status
C. Spiral Wound	1. inexpensive ($\sim 3\$/ft^2$) 2. very compact with low holding volume per unit membrane area 3. well-developed equipment 4. low parasitic pressure drop losses	1. very susceptible to plugging 2. very difficult to clean and probably not acceptable for food, dairy and pharmaceutical applications	currently available for RO only, but potentially adaptable to UF
Laminar Flow			
A. Flat Channels; Annuli, etc.	1. potentially more economical for highly viscous solutions 2. very low holding volume per unit membrane area 3. high conversion per pass obtainable; eliminates need for recycle operation 4. low parasitic pressure drop losses	1. equipment not well-developed 2. susceptible to plugging 3. potentially difficult to clean 4. presently expensive 5. membrane replacement difficult (depends on exact design)	large-scale equipment under development

Table III (Continued)

MEMBRANE CONFIGURATIONS FOR ULTRAFILTRATION

System	Advantages	Disadvantages	Commercial Status
B. Hollow Fibers (tubeside feed)	1. same as above, plus potentially inexpensive	1. not developed 2. susceptible to plugging 3. potentially difficult to clean 4. entire fiber cartridge must be replaced upon single fiber failure	not available
C. Hollow Fibers (shellside feed)	1. same as above, plus potentially inexpensive	1. excessive plugging excludes use for most UF applications 2. non-cleanability excludes use for food, dairy and pharmaceutical applications	commercially available for RO only

* parallel flat leaves attached to a central permeate collection header

over 1 billion pounds of lactose and more than 150
million pounds of soluble milk proteins. Both can be of
substantial dollar value if recovered. A demonstration
grant by the Federal Water Quality Administration has
been made to Crowley's Milk Company, Binghamton, New
York, to install a membrane process, primarily to reduce
the water pollution arising from the discharge of cottage
cheese whey, and secondarily to recover by-products to
defray the costs of the membrane process. Under this
grant Abcor, Inc., has been the subcontractor, responsible
for process development, and plant design and construction.

Figure 3 shows a simplified flow sheet for the two-
step process. Cottage cheese whey is fed to a low pressure ultrafiltration unit for whey protein recovery and
concentration. The permeate from the ultrafiltration
section is concentrated by reverse osmosis for lactose
recovery (e.g., by crystallization). The bulk of the
water contained in the whey is passed through the reverse
osmosis membranes and recovered as a low BOD product
water. Operation of the ultrafiltration section is of
interest here, and only data for it will be presented.

Both non-protein nitrogen and true protein retentions
have been determined by gel permeation chromatography
for Abcor HFA-180 tubular membranes, Figure 4. Low molecular-weight nitrogen-containing material passes the membrane while protein retention is almost quantitative ($\approx$99).
Retention of other low molecular weight solutes (minerals,
lactose, lactic acid) is nil, as determined by analysis.

The rate of passage of solution through the membrane is
a critical design parameter because it determines the
membrane area requirement, or plant size, and thus capital
cost. Major factors affecting flux are flow rate through
the membrane unit, operating temperature, and feed concentration level.

Figure 5 shows the influence of these variables.
Flux typically decreases with increasing feed protein concentration, as well as with decreasing temperature. For
high flux, the optimum operating temperature appears to be
about 125°F. Neither the membranes nor the whey proteins
tolerate temperatures much above this level. For cottage
cheese whey this also minimizes plant heat exchange requirements since whey is received hot from the vats.
Preliminary results show little heat denaturation of the
proteins at this temperature. Furthermore, the rate of

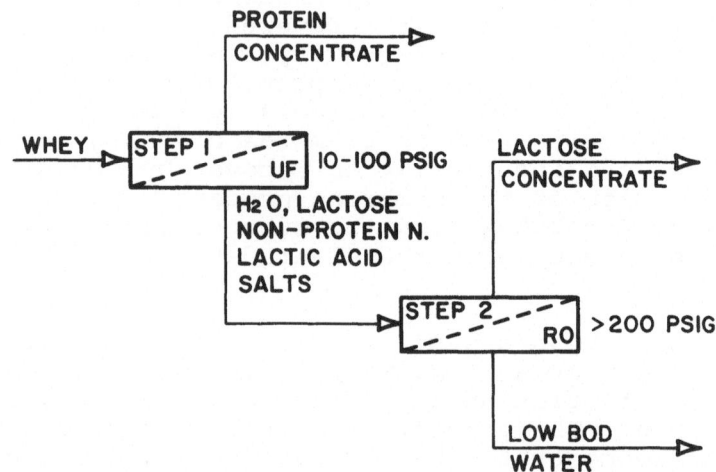

FIGURE 3 -MEMBRANE PROCESS FOR WHEY
TREATMENT--FLOW SCHEMATIC

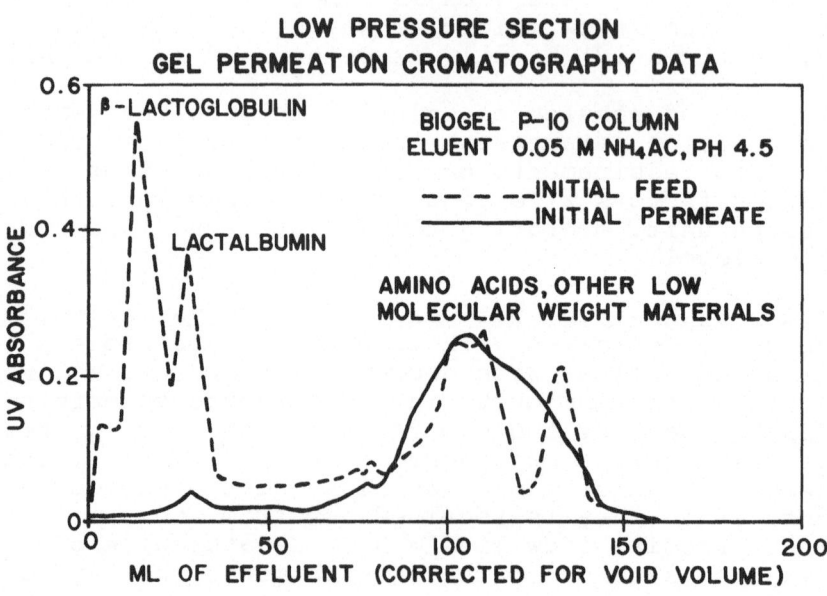

FIGURE 4

LOW PRESSURE SECTION
GEL PERMEATION CROMATOGRAPHY DATA

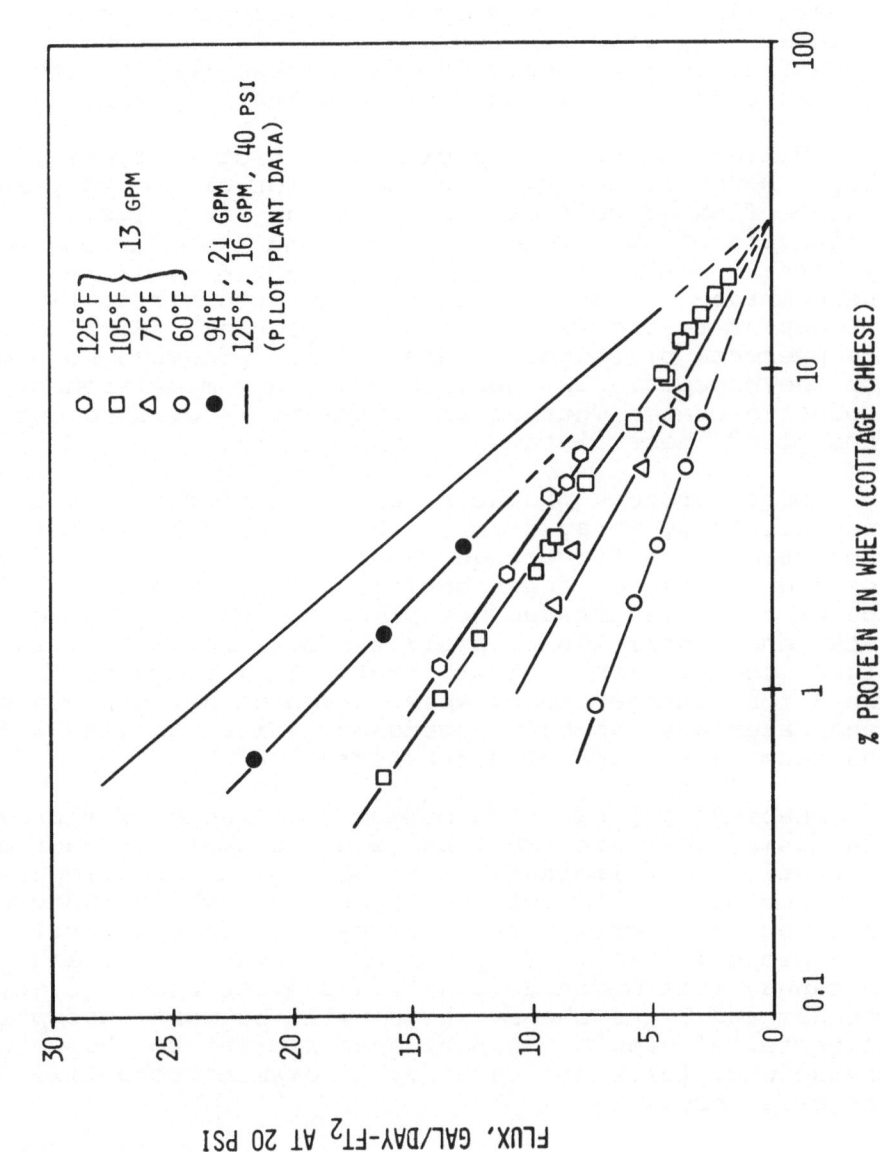

FIGURE 5: ULTRAFILTRATION FLUX AS FUNCTION OF TEMPERATURE AND WHEY PROTEIN CONCENTRATION

bacterial growth at 125°F appears to be very low, at least in acid wheys.

The data for the Crowley's 1000 lbs/hour pilot plant show high flux levels. Additional operating data show that flux increases almost in proportion with the linear velocity of the whey in the membrane system (suggested by data in Figure 5), and is independent of operating pressure above 15 psi. These results are consistent with flux limited by concentration polarization.

Ultrafiltration is applicable to other types of whey. However, the type and condition of the whey influence flux as well as the operating variables. One limitation of the process, is that no free fat can be present. Although a separate fat phase can coat the membrane, greatly reducing flux, emulsified fat does not appear to create a processing problem. Similarly, the presence of residual casein fines presents no problem for the process. The desired protein composition of the product dictates whether pretreatment is carried out for removal of these fines.

Dried protein products can be recovered from the concentrate by spray drying. Pilot tests have yielded products from both cottage and cheddar cheese wheys with no evident loss of functionality. The composition of the protein product depends primarily on the degree of water and impurities removal; the more water removed, the higher the protein content. This is illustrated in Figure 6 for cottage cheese whey. Even higher protein content materials can be prepared by rediluting with water and reconcentrating by ultrafiltration.

Cheese whey can be a medium for growth of microorganisms, and care must be taken to avoid product contamination. Preliminary micro-biological experiments have been performed with cottage cheese whey which indicate that bacterial growth problems can be satisfactorily handled in a properly designed system[14]. Because of the higher pH range, intermediate acidity and sweet wheys (e.g., cheddar and Swiss cheese wheys) will be more susceptible to bacterial growth. Experiments similar to the cottage cheese whey tests are underway to examine potential bacterial problems.

System cleaning and sanitation techniques are also being examined. Of concern is the potential degradation of membranes by strong chemical agents normally used in food

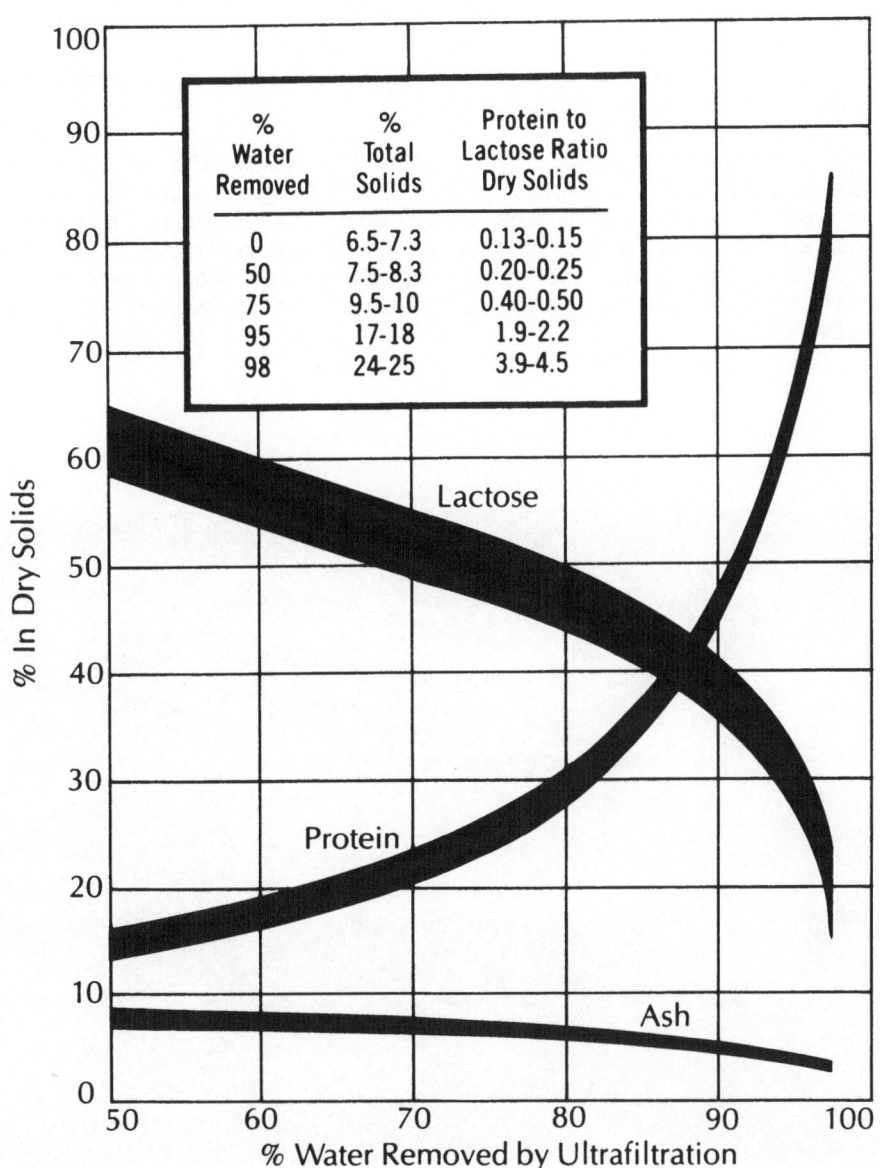

% Water Removed	% Total Solids	Protein to Lactose Ratio Dry Solids
0	6.5-7.3	0.13-0.15
50	7.5-8.3	0.20-0.25
75	9.5-10	0.40-0.50
95	17-18	1.9-2.2
98	24-25	3.9-4.5

FIGURE 6: COMPOSITION OF WHEY PROTEIN CONCENTRATE

and dairy systems. Initial results show, however, that
certain detergents as cleaners, and free chlorine as a
sanitizer, do not degrade the ultrafiltration membranes.[14]
These have been used on a regular basis in a membrane
life test with operation and membrane life currently exceeding one year.

At present, protein recovery through ultrafiltration
is being studied in large pilot plant operations. A
photograph of the 1000 lbs/hour ultrafiltration system
installed in the Crowley's Milk Company pilot plant as
part of the F.W.Q.A. demonstration grant is shown
in Figure 7. Operation of this unit at Crowley's began
in July, 1970.

FIGURE 7: ULTRAFILTRATION MODULE AT CROWLEY'S MILK COMPANY FOR COTTAGE CHEESE WHEY FRACTIONATION

Based on current data, capital and operating costs have been projected for sanitary-design installations of varying whey treatment capacity.[16] Figures 8 and 9 show these for ultrafiltration, and a comparison with direct concentration by reverse osmosis and vacuum evaporation. The more favorable economics of the membrane processes are evident. The incentive for ultrafiltering whey to produce a protein product is very large. Because of the high quality of the protein produced in the process, it should command a price substantially in excess of 50¢/lb. Since operating costs are low, this provides the cheese plant operator with a high return on the investment required to install the process, a new product opportunity, and, when UF and RO are combined, an answer to the problem of pollution from whey.

Electrocoating Paint

Electrocoating is a recently developed process for applying a very uniform, controlled layer of paint on metal (mainly primers), from a colloidal, aqueous paint suspension. The parts travel through a dip tank (Figure 10) and the paint coating is applied under an electric potential. As the resin particle is deposited, it loses its charge and becomes water insoluble. The paint resins are chemically solubilized (with amine or KOH) and as the resin is removed from the tank in the form of coating on a part, the tank paint becomes progressively more dilute in resin particles (pigment). In addition, an electrocoating paint solution is in a rather delicate chemical balance, and can be seriously harmed by introducing certain ionic species (such as phosphates) which may be carried into the bath from previous processing steps.

After painted parts leave the coating tank, the excess paint, called "drag-out", is rinsed off before the paint is heat-cured. Currently, this rinse is wasted, at an economic loss. It also causes a substantial pollution problem.

There are three basic uses of ultrafiltration in the electrocoating process. All relate to the fact that ultrafiltration membranes retain resin particles but pass essentially all the dissolved lower molecular weight solutes (salts, stabilizers, etc.). These are, first to remove excess solubilizing agents and foreign ions from the bath, thus increasing operational reliability ("tank balancing"); second, to recover the paint pigment lost as "drag-out"; and third, to eliminate pollution and reduce the deionized rinse water requirement.

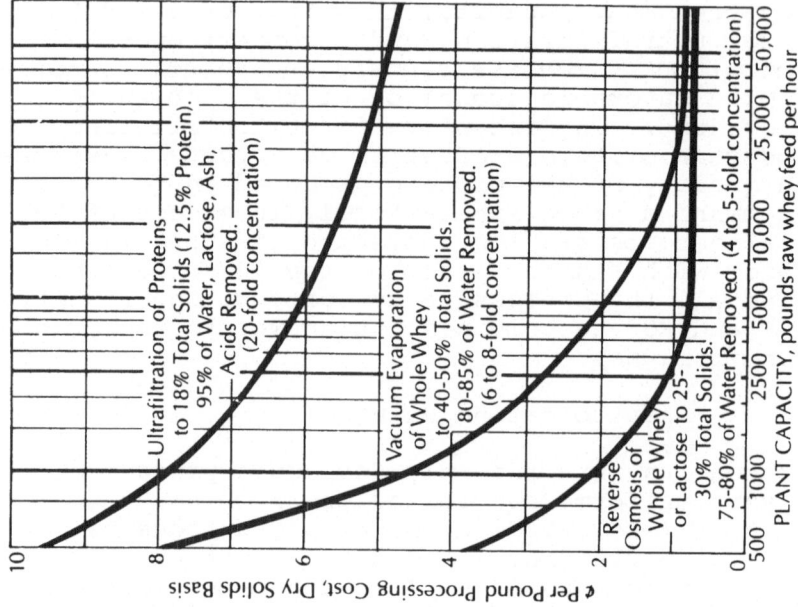

FIGURE 9: WHEY TREATMENT OPERATING COSTS

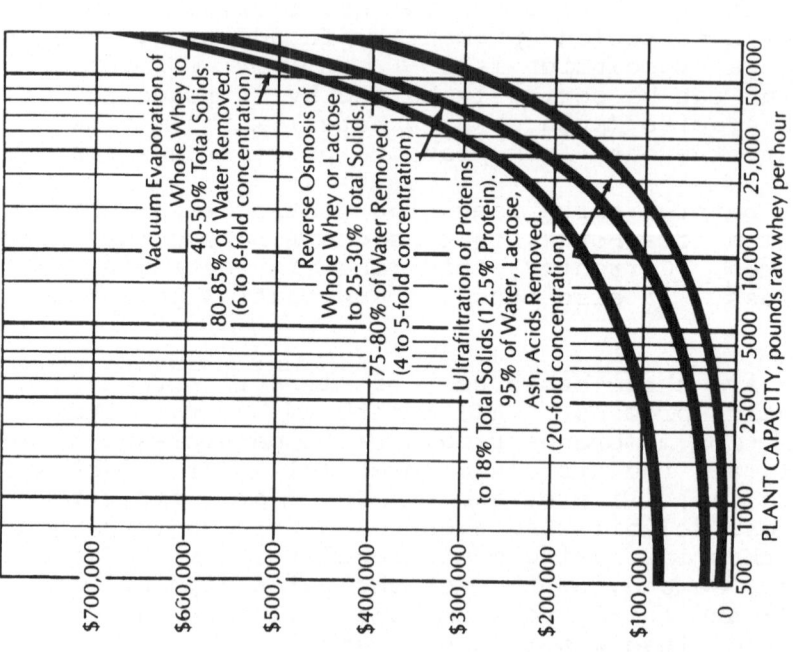

FIGURE 8: WHEY TREATMENT CAPITAL COSTS

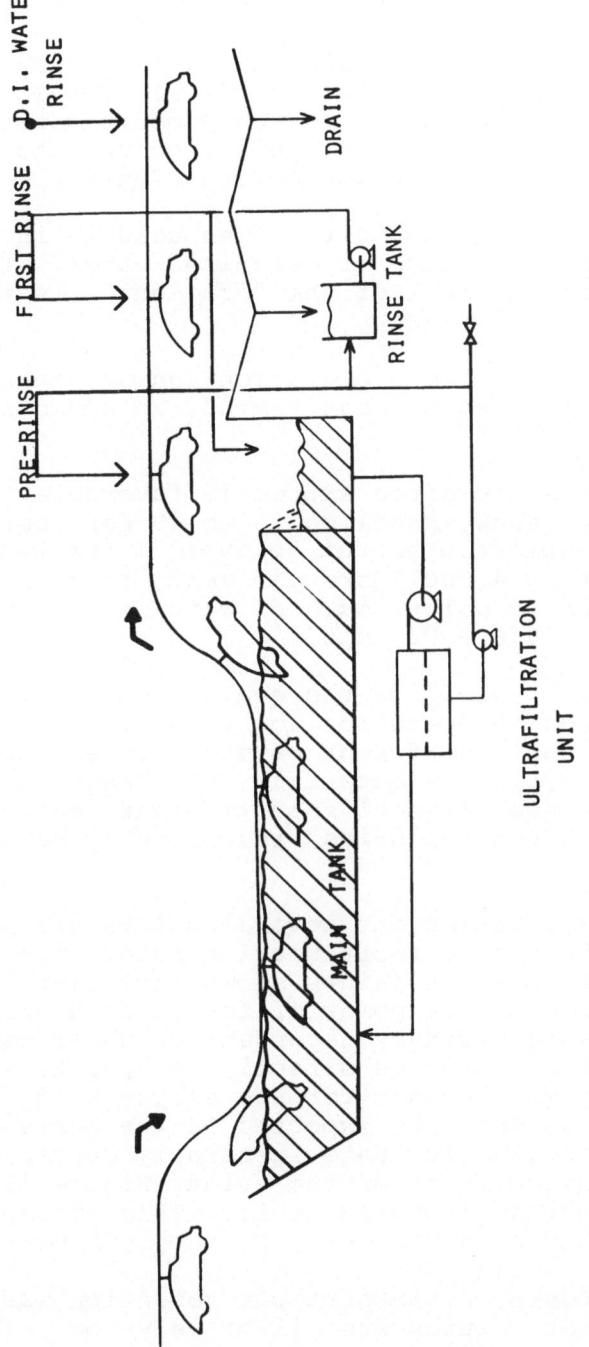

FIGURE 10: SCHEMATIC OF ELECTROCOATING INSTALLATION WITH ULTRAFILTRATION UNIT

A very effective means of accomplishing these needs
in a single step is to install an ultrafiltration unit to
treat the tank contents. Ordinarily, the paint is
recirculated through the tank in order to maintain uniform
tank composition. A differential ultrafiltration unit
can readily be installed in this recirculation loop. For
new electrocoating processes, the only modification may
be to use a slightly larger recirculation pump to account
for frictional head loss in the membrane unit.

The permeate from the unit can be used in initial or
all rinsing steps, in place of deionized water. This
rinse, containing the bulk of the "drag-out", is returned
directly to the paint tank.

This relatively simple operation can substantially
reduce the deionized water requirement, an added economic
attraction.

The economics are often extremely favorable. In an
automotive "body" tank, handling 30 to 40 cars per hour,
the paint "drag-out" can amount to over $1 per car or $30-
40/hour. Operating 4,000 hours per year, this is $120,000
to $160,000 worth of paint, much of which can be saved by
installing a $25-50,000 UF unit.

One pilot unit used for the evaluation of different
materials has been in operation for about six months.
While treating twelve different paints, representative
of commercially available materials, no change in flux or
membrane performance (pigment retention)has been observed.
On this basis, industrial units are currently being installed.

In Figure 11, fluxes for several paints are shown as
a function of the system recirculation rate. The data
presented for these eight paints shows that flux increases
with about the 1.0 to 1.1 power of the recirculation rate.
As the viscosity is strongly dependent on shear rate, the
recirculation rate cannot be directly related to Reynold's
number, and this may account for the slopes being higher
than about 0.9, as would be expected on the basis of
Equation 2. Although flux rates generally decrease with
increasing solids contents of the paint, Figure 12, this
"rule" has exceptions (note paint B). This probably
reflects variations in viscosity of the different samples.

For these tests, resin particle retention was 100%.
Retention of other solutes was nil or very low. In

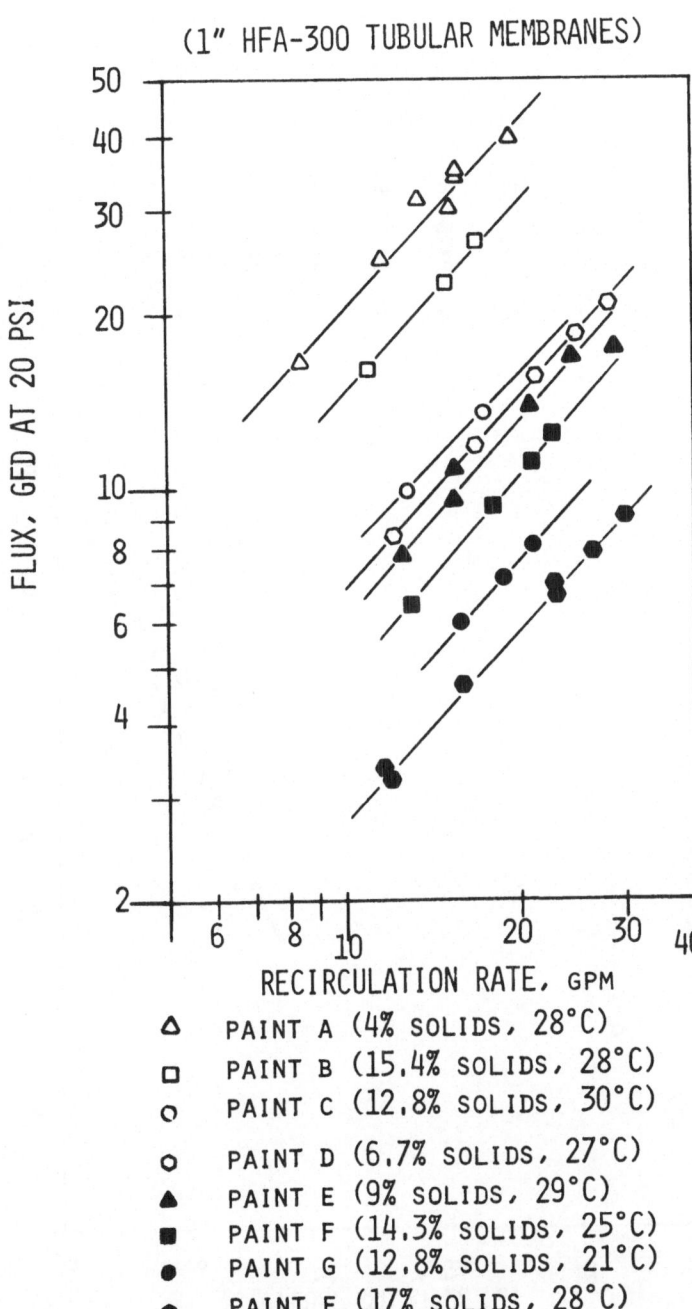

FIGURE 11: PAINT ULTRAFILTRATION

(1" HFA-300 TUBULAR MEMBRANES)

△ PAINT A (4% SOLIDS, 28°C)
□ PAINT B (15.4% SOLIDS, 28°C)
○ PAINT C (12.8% SOLIDS, 30°C)

⬠ PAINT D (6.7% SOLIDS, 27°C)
▲ PAINT E (9% SOLIDS, 29°C)
■ PAINT F (14.3% SOLIDS, 25°C)
● PAINT G (12.8% SOLIDS, 21°C)
⬢ PAINT E (17% SOLIDS, 28°C)

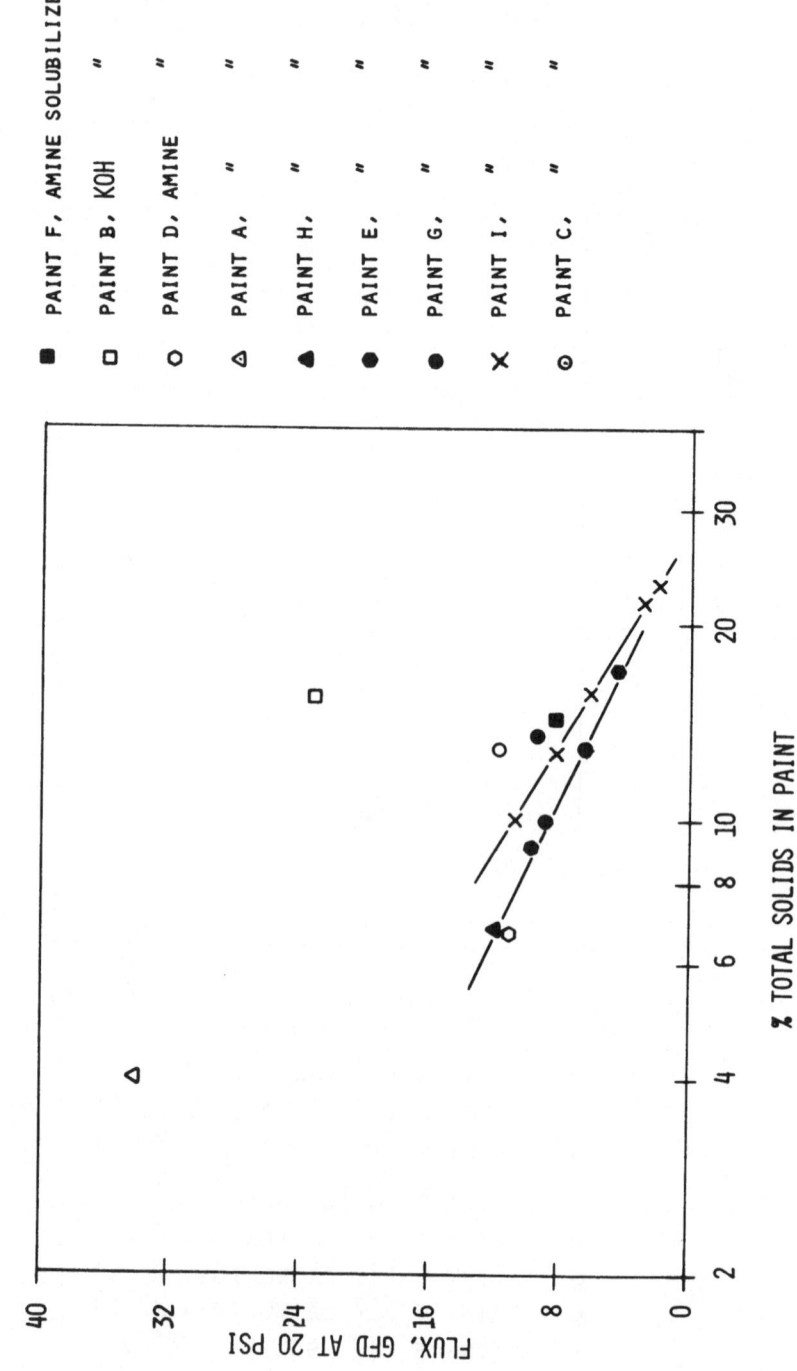

FIGURE 12: PAINT ULTRAFILTRATION (1" HFA-300 TUBULAR MEMBRANES; 16 GPM FEED FLOW; 27°C)

■ PAINT F, AMINE SOLUBILIZED
□ PAINT B, KOH   "
○ PAINT D, AMINE   "
△ PAINT A,   "   "
◆ PAINT H,   "   "
● PAINT E,   "   "
● PAINT G,   "   "
✕ PAINT I,   "   "
⊙ PAINT C,   "   "

certain cases, ionic solutes "associated" with the
resin particles by electrostatic forces were partly
retained by the membrane.

The electrocoating application is an excellent one
for ultrafiltration. Economics are highly favorable and
the process can be easily integrated into both new and
existing paint lines.

Enzymes

The large-scale production and isolation of industrial enzymes has received major impetus from their
introduction into laundry detergents, and other large
market applications. In addition, long-term forecasts
predict substantial growth in the production and use of
enzymes.

Enzymes are classically generated by fermentation
(microbial enzymes) or extraction (animal or vegetable
enzymes) in aqueous media. Purification and recovery in
dry form are frequently the most troublesome steps in
the production process.   Recent reviews describe problems
and limitations of these operations - usually relating to
their high cost or product denaturation (loss of enzymatic activity).(2,3,9)

Following separation of suspended solids, including
cell debris, the enzyme broth can be treated by several
means. Purifications include salt precipitation (e.g.,
with ammonium sulfate), solvent precipitation (e.g.,
ethanol), ion exchange, crystallization, gel permeation
chromatography, dialysis, etc. Final recovery in solid
form is frequently effected through spray drying, often
preceded by a concentration step, e.g., by vacuum evaporation.

Recently, ultrafiltration has been shown to be a
very effective means of purifying enzyme broths while simultaneously concentrating them. Figure 13 shows a typical flow schematic. During concentration lower molecular
weight solutes (salts, sugars, other materials) are
passed through the membrane, thus enhancing the "specific
activity" of the final product. This may not always be
desirable, since some solids, in particular, salts, can
contribute to enzyme activity and/or stability.

In addition to increasing the enzyme content on a
dry basis, more subjective purification can be achieved.
This applies primarily to color and odor removal, which

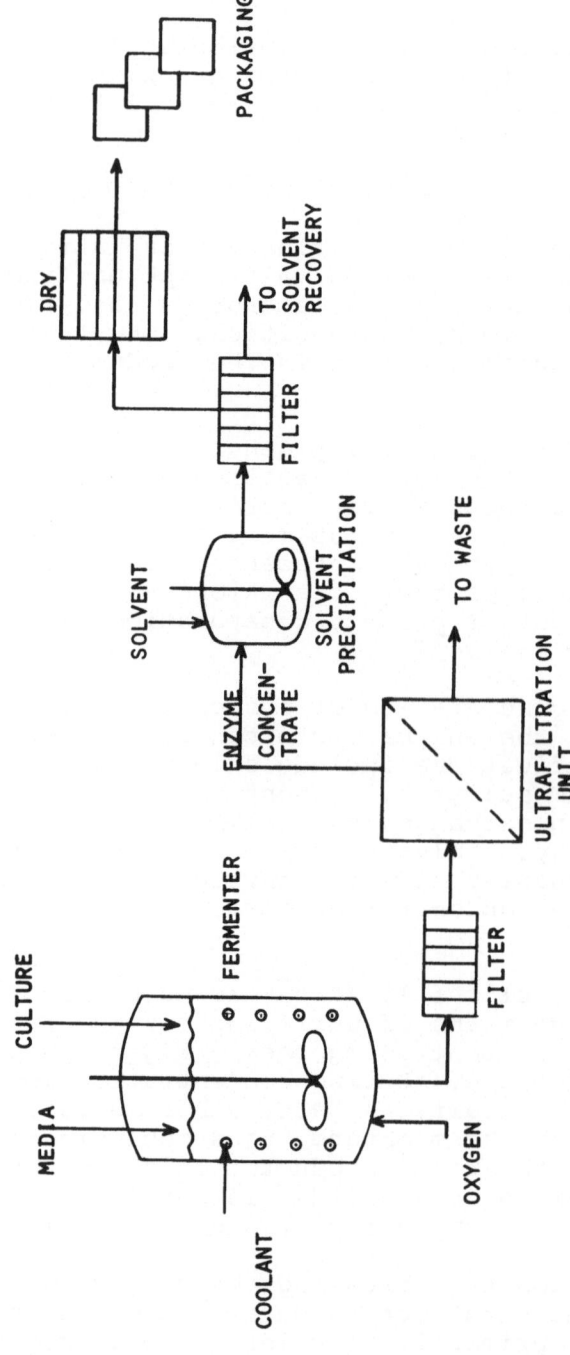

FIGURE 13: SCHEMATIC OF ENZYME PRODUCTION INSTALLATION WITH ULTRAFILTRATION UNIT

contributes to increased product value. By contrast,
the common alternative means of concentrating enzymes -
vacuum evaporation - sometimes leads to intensification
of odor and color by chemical reaction.

As ultrafiltration can be performed at room temperature and even down to below 2°C, loss in activity due to
exposure to high temperature is avoided, and in the concentration of proteases, "auto-digestion" can be minimized.

Finally, high enzyme concentration ratios can be
achieved without major increase in solution viscosity,
since un ultrafiltration, lower molecular weight materials are not retained by the membrane. Included
are sugars which especially contribute to a rapidly
increasing viscosity with concentration, frequently limiting concentration by evaporation.

A wide variety of enzyme broths have been concentrated in our laboratories. Initial tests have examined
membranes to determine enzyme retention and purification
properties. Having chosen appropriate membranes, pilot
tests were conducted to examine flux as a function of
concentration ratio and system operating parameters, and
enzyme yield. Even with 100% enzyme retention, yield is
generally less, due to some loss in activity. This is
believed to arise primarily from thermal denaturation
and/or shear denaturation in various parts of the non-
membrane system (pump, pressure control valve, rotameter,
etc.).   In most cases, however, enzyme activity accounta-
bility has exceeded 90%.

Flux data for four enzyme broths are shown in Figure
14. Data are plotted against concentration ratio, rather
than enzyme concentration as this information was not
obtained in all cases. Differences in flux levels relate
primarily to differences in enzyme concentration (gm/
liter), recirculation rate (gpm), and temperature. As
expected, flux decreases with increasing concentration
and increases with increasing recirculation rate and
temperature.

Three enzymes were produced by fermentation; the
veal rennet was extracted. Broths were centrifuged or
filtered for suspended solids removal prior to concentration, although this is not required for acceptable performance of the ultrafiltration equipment.

Other enzymes which have been examined include cellu-
lase, glucose oxidase, trypsin, lysozyme, pepsin,

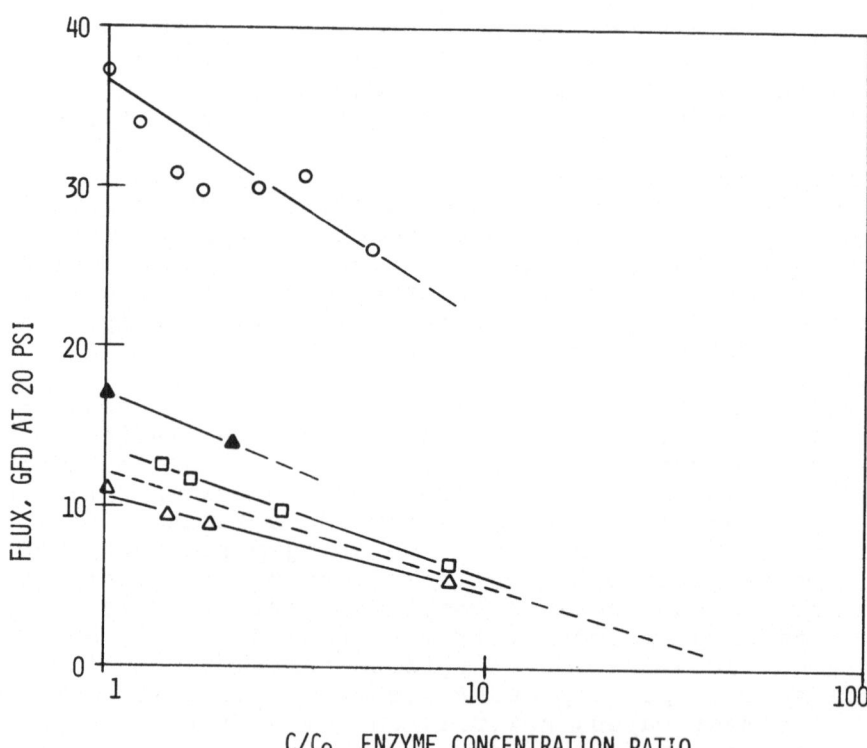

FIGURE 14:   ULTRAFILTRATION OF ENZYME BROTHS

O 50,000 MW PROTEASE, 1 INCH HFA-200 TUBULAR MEMBRANE;
   13 GPM FEED FLOW ; 28°C.

▲ AMYLASE/PROTEASE MIXTURE, 4.5% INITIAL SOLIDS; 1 INCH
   HFA-300 TUBULAR MEMBRANE; 10 GPM FEED FLOW; 28°C.

□ VEAL RENNET EXTRACT; 1 INCH HFA-300 TUBULAR MEMBRANE;
   10 GPM FEED FLOW, 32°C.

--- MICROBIAL ENZYME; LINE DRAWN THROUGH 15 DATA POINTS TO
    35X CONCENTRATION; 1 INCH HFA-200 TUBULAR MEMBRANE; 20
    GPM FEED FLOW; 30°C.

△ MICROBIAL ENZYME; HFA-300 SHEET MEMBRANE IN LAMINAR FLOW
   CELL WITH 60 MILS SPACING (RE ≃1000); 2.6 GPM FEED FLOW
   15°C.

α-chymotypsinogen, and insulin in appropriate extracts
or buffered solutions.

Miscellaneous Applications

The three applications of ultrafiltration described
above are typical of current uses for this process.
There are currently several others being practiced on an
industrial scale. Many other applications are under development and as this unit operation becomes more widely
known, there will be many areas where it becomes the
processing technique of choice. Examples of applications
under development in our laboratories now are

-- skim milk protein concentration
-- soy whey protein fractionation and concentration
-- serum protein recovery from slaughterhouse wastes
-- yeast processing wastes
-- vegetable processing wastes
-- gelatin concentration
-- fetal calf serum ultrafiltration for protein purification and recovery (for continuous mammalian
   tissue culture plant)
-- virus concentration
-- vaccine concentration and purification
-- chemical plant effluent containing higher MW alcohols and plasticizers
-- industrial dye suspensions
-- metal plating wastes
-- acrylic latices
-- pulp and paper wastes

SUMMARY

This paper has examined some of the operational and
theoretical considerations pertinent to the industrial
applications of ultrafiltration. In practice, membranes
are chosen experimentally to have the desired fractionation (rejection) properties. Flux is almost always limited by the phenomenon known as concentration polarization. The major problem in ultrafiltration system design
is to minimize this effect by maximizing the mass transfer of retained solute away from the membrane surface.
In laminar and turbulent flow systems, this is accomplished by flowing the solution past the membrane at a
high velocity, generating high wall shear rates.

Three major applications have been discussed: whey
fractionation and concentration, the treatment of elec-

trophoretic paint suspensions, and enzyme purification and concentration. These three currently are being practiced industrially. Several other applications have been mentioned.

Process design for new high conversion systems is generally performed on a case by case basis utilizing the type of approach described in this paper together with experimental data. Important considerations are costs for the membrane modules, pumps, and associated controls and equipment; and also operating costs for power, chemicals, etc.

REFERENCES

(1)   Baker, R.W., and Strathmann, H.,"Ultrafiltration of Macromolecular Solutions with High-Flux Membranes", J. App. Poly. Sci., 14, 1197(1970)

(2)   Beckhorn, E.J., "Production of Industrial Enzymes", Wallerstein Laboratories Communications, Vol. XXIII, No. 82, Dec.(1960)

(3)   de Becze, G.I., "Enzymes,Industrial", in Kirth-Othmer Encyclopedia of Chemical Technology, Vol. 8, 2nd Ed., 173(1965)

(4)   Bixler, H.J., Hausslein, R.W., and Nelsen, L., "Separation and Purification of Biological Materials by Ultrafiltration", presented at the 65th Nat. AICHE Meeting, Cleveland, May (1969)

(5)   Blatt, W.F.,Dravid, A., Michaels, A.S., and Nelsen, L., "Cake Formation in Membrane Ultrafiltration", in Membrane Science and Technology (J.E. Flinn, Ed.), Plenum Press, New York (1970)

(6)   Brian, P.L.T., "Mass Transport in Reverse Osmosis", in Desalination by Reverse Osmosis (U. Merter, Ed.), The M.I.T. Press, Cambridge (1966)

(7)   Brown, C.E., Tulin, M.P., and Van Dyke, P., "On the Gelling of High Molecular Weight Impermeable Solutes During Ultrafiltration", presented at the Third Joint Meeting Institute de Ingenieros Quimicas de Puerto Rico and AICHE, San Juan, P.R., May (1970)

(8)   deFilippi, R.P., and Goldsmith, R.L., "Application
      and Theory of Membrane Processes for Biological
      and other Macromolecular Solutions", in Membrane
      Science and Technology (J.E. Flinn, Ed.), Plenum
      Press, New York (1970)

(9)   Dunnill, P., and Lilly, M.D., "Large Scale Isolation
      of Enzymes", Process Biochemistry, page 13, July
      (1967)

(10) Fenton-May, R.I., Hill, C.G., and Amundson, C.A.,
      "Use of Reverse Osmosis/Ultrafiltration Systems
      for the Concentration and Fractionation of Whey",
      Dept. of Food Science, Univ. of Wisconsin,
      private communication

(11) Flory, P.J., Principles of Polymer Chemistry,
      Cornell University Press, Ithaca (1957)

(12) Ginnette, L.F., and Merson, R.L., "Maximum Permea-
      tion Rate in Reverse Osmosis Concentration of
      Viscous Materials", presented at the 73rd Nat.
      AICHE Meeting, St. Louis, Feb. (1968)

(13) Goldsmith, R.L., "Macromolecular Ultrafiltration
      with Microporous Membranes", I&EC Fundamentals
      Quarterly, in press

(14) Goldsmith, R.L., Amundson, C.A., Horton, B.S., and
      Tannenbaum, S.R., "Recovery of Cheese Whey
      Proteins through Ultrafiltration", presented at
      SOS/70, Washington, D.C., August (1970)

(15) Harriott, P., and Hamilton, R.M., "Solid-liquid
      Mass Transfer in Turbulent Pipe Flow", Chem. Eng.
      Sci., 20, 1073 (1965)

(16) Horton, B.S., Goldsmith, R.L., Hossain, S., and
      Zall, R., "Membrane Separation Processes for the
      Abatement of Pollution from Cottage Cheese Whey",
      presented at the Cottage Cheese and Cultured Milk
      Products Symposium, Univ. of Maryland, March (1970)

(17) Michaels, A.S., "New Separation Technique for the
      CPI", Chem. Eng. Prog., 64, No. 12, 31(1968)

(18) Michaels, A.S., "Ultrafiltration", in Advances in
      Separations and Purifications, (E.S. Perry, Ed.),
      p. 297, J. Wiley and Sons, New York (1968)

(19) Peri, C., and Dunkley, W.L., "Influence of Composi-
      tion of Feed on Reverse Osmosis of Whey", presented
      at SOS/70, Washington, D.C., March (1970)

(20) Wang, D.I.C., and Butterworth, T.A., Dept. of Food
      Science and Nutrition, M.I.T., private communica-
      tion, Sept.,(1970)

SOME ASYMMETRY PROPERTIES OF COMPOSITE MEMBRANES

N. Lakshminarayanaiah and Fasih A. Siddiqi

Department of Pharmacology, University of Pennsylvania

School of Medicine, Philadelphia, Pa. 19104.

ABSTRACT

Composite membranes of two types, layer-type and sandwich-type, have been prepared from 2% collodion solutions containing different amounts of polystyrene sulfonic acid. The layer-type membranes were formed by evaporating first a collodion solution containing the least amount of the polyelectrolyte. On top of this collodion solutions containing increasing quantities of the polyelectrolyte were evaporated one after the other. Sandwich-type membranes were formed by trapping some polyelectrolyte between two simple membranes, one containing a small quantity and the other containing a large quantity of the polyelectrolyte.

Electrical potentials arising across these membranes when they separate the same concentration of (1:1) electrolyte have been measured as a function of concentration. Long term equilibration of these membranes with very dilute electrolyte solutions tend to lower the potentials observed in the first few hours of equilibration. These results are discussed in terms of a model system built from two simple membrane cells, one containing membrane of high charge density and the other containing membrane of low charge density, in such a way that one membrane cell opposed the emf of the other membrane cell.

INTRODUCTION

A group of Italian workers[1-3] have shown that composite membranes prepared from collodion-polystyrene sulfonic acid can mimic some of the properties of the squid nerve membrane. In fact one of them[4] has advanced a physico-chemical model for the nerve

301

membrane in terms of which the resting and action potentials have
been explained. These studies intrigued us very much and we repeated
their work. Short period (1-2 hr) equilibrations of the membrane
with the electrolyte solutions, particularly the layer-type
membranes, gave values for the asymmetry potential[3] (i.e. emf
generated when the same concentration of the electrolyte was placed
on either side of the membrane) which agreed roughly in magnitude
with the published values of the Italian workers[2]. In this paper,
the results of long term equilibration and of other experiments
are presented.

EXPERIMENTAL

The dissolution method reviewed by Neihof[6] was used to prepare
simple membranes of collodion which contained various amounts of
polystyrene sulfonic acid (PSSA). Styrene polymer (Borden Chemical
Co., Philadelphia, Pa) was sulfonated and PSSA obtained following
the procedure described by Neihof. Membranes were cast on clean
and dry glass plates from a 2% solution of Parlodion (purified
nitrocellulose, Mallinckrodt) in alcohol-ether mixture (3:1 by
volume) containing different amounts of PSSA.

The composite membranes were prepared following the steps
given by Liquori and Botre[1,2], although the compositions of solutions used to form membranes were different. The layer-type membranes
were formed by casting different layers of 2% Parlodion solution
containing various amounts of PSSA (8.3 to 0.02 mg/ml), one on top
of the other, in the way described by Liquori and Botre[1]. The
sandwich-type membranes were formed by trapping a layer of PSSA
(4 mg/ml) between high charge density membrane (2% collodion, USP,
Fisher, containing 10 mg/ml PSSA) and low charge density membrane
(2% collodion, USP, Fisher, containing 1 mg/ml PSSA) in the way
outlined by Liquori and Botre[2]. In this case alcohol to ether ratio
was 1:2.5.

The membranes were converted into the Na form by treating them
with 1.0 N solution of NaCl. They were washed with deionized water
and equilibrated with the solution to be used in the experiment
with frequent changes of the solution. The asymmetry potentials
arising across composite membranes were measured on a Keithley
microvoltmeter in an air conditioned room at 22°C using an all
glass H-cell and saturated KCl-agar bridges.

RESULTS

Electrical potentials (high charge density side taken positive)
arising across some of the composite membranes, both layer and
sandwich types, when they were subject to short term equilibrations
(1-2 hr) and separating the same electrolyte solution of NaCl are
given in Figure 1. The membranes were screened to choose only the

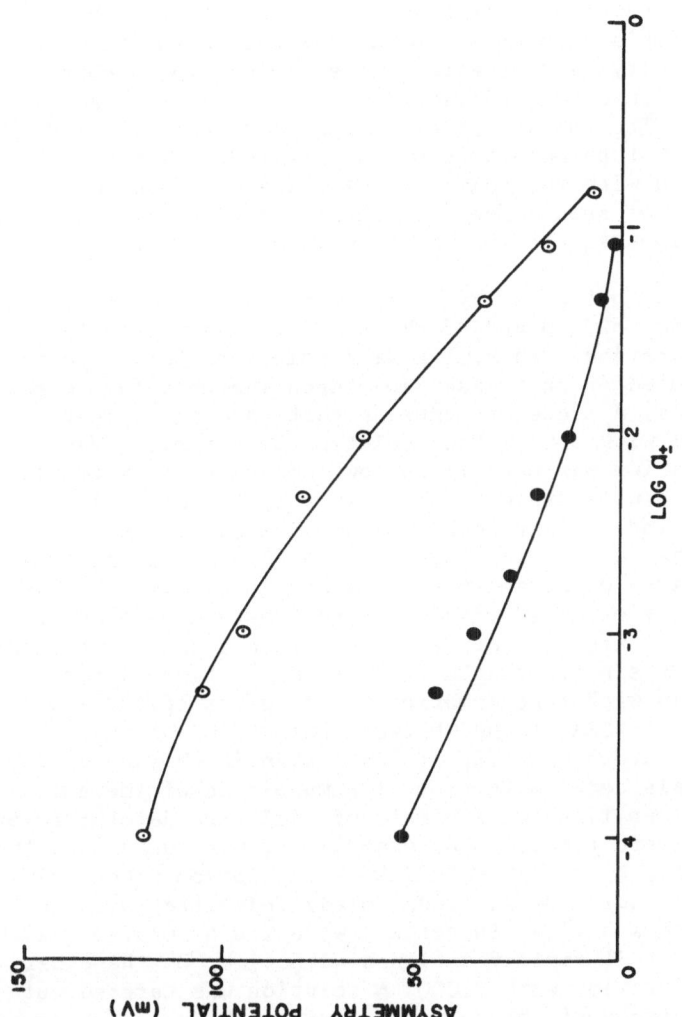

Fig. 1. Electrical potentials arising across a composite membrane when it separates the same concentration of NaCl plotted as a function of the activity of the external solution. (⊙) refer to asymmetry potentials across a layer-type membrane; (●) refer to asymmetry potentials across a sandwich-type membrane.

majority of those which gave good and steady emf values agreeing
among themselves within ± 10%. There were however some membranes
belonging to the same lot which gave emf values lower in magnitude
to those given in Figure 1. The sandwich-type membranes with
which we had the most difficulty in preparing them, gave asymmetry
potentials which were lower than those given by layer-type
membranes (see Figure 1) and those observed by Liquori and Botre[2]
for similar membranes (see their Figure 3). Deposition of PSSA
between the two simple membranes, one of low charge and the other
of high charge density, always gave good sandwich-type membranes;
but they separated into the individual units when kept in equili-
brating solutions. The results given in Figure 1 were obtained
from those sandwich-type membranes which remained intact
after equilibration with the solutions. Inspite of repeated
attempts, we were not successful in preparing sandwich-type membranes
which gave emf values as high as those published by Liquori and
Botre[2].

     Four pieces cut out of a layer-type membrane formed on a
glass plate gave 62, 61, 50 and 45 mV each for asymmetry potential
after 2 hr equilibration with 0.01 N NaCl solution. This type of
variability was noted in most cases and hence the need for screen-
ing these membranes. A piece of membrane that gave an asymmetry
potential of 54 mV with 0.01 N NaCl solution in the first few
hours of equilibration was used to follow the change in potential
with time. These results are given in Figure 2. The potential
decayed with time and similar decay was noted also at other
concentrations. The decline was usually slow in the case of those
membranes which gave low asymmetry potentials initially. Similar
pattern was commonly observed with our sandwich-type membranes.
Included in Figure 2 are the data realized using a genuine Liquori-
Botre membrane (our sincere thanks to Prof. A. M. Liquori for
supplying their sandwich-type membrane to one of us (F.A.S)) which
contained a layer of PSSA trapped between two collodion films
containing 0.5 and 0.005 equiv/Kg of PSSA. Even in this case, the
asymmetry potentials decay with time. The magnitude of the emf
observed at any given time and its rate of fall were determined by
the way membranes were treated. For example, it was found that the
potential was fairly high with 0.0001 N NaCl solution (about 120-
130 mV) when the measurement was made immediately after washing the
Na form of the membrane (i.e. in contact with 1.0 N solution) with
deionized water. But it rapidly declined with time. On the other
hand if the equilibration with 0.0001 N solution was carried out
for 10-12 hr with frequent changes of solution, the value observed
for the potential was low. Again the strength of the solution used
in the measurement also affected the magnitude of the potential.
Generally low values were observed with solutions whose concentration
was greater than 0.01 N. The factors causing this behavior are
discussed below.

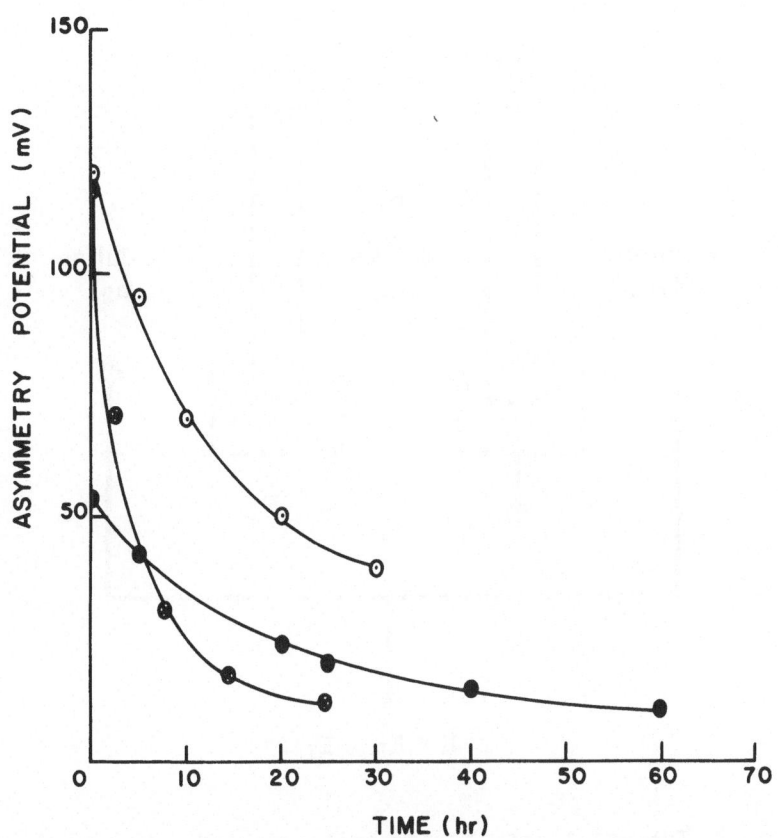

Fig. 2. The decline of asymmetry potential followed as a function of time. (☉), (●) refer to a layer-type membrane in contact with 0.0001 and 0.01 N NaCl solution respectively. (⊛) refer to a genuine Liquori-Botre sandwich-type membrane in contact with 0.0001 N NaCl solution.

DISCUSSION

The asymmetry potential was considered by Liquori and Botre in their earlier paper[1] in terms of Donnan and diffusion potentials. Later[2] they proposed a model (see Figure 3) in which two membranes, one of high charge density M_H (i.e. highly selective to cations) and the other of low charge density M_L (i.e. low selectivity to

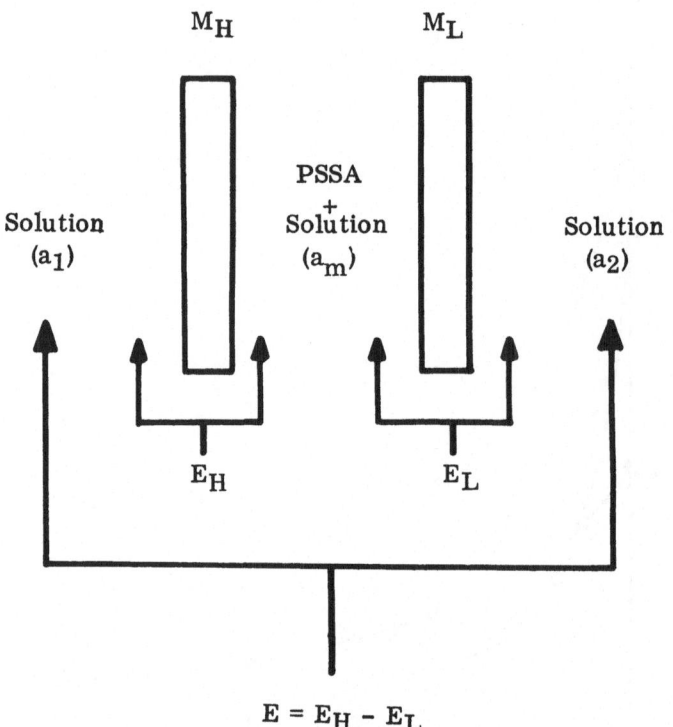

Fig. 3. The model for the composite membrane. M_H and M_L are simple membranes of high charge (PSSA = 2.5 mg/ml of membrane forming solution) and low charge (PSSA = 0.025 mg/ml) density respectively separating three solutions of the same electrolyte of activity a_1, a_m and a_2 contained in the three chambers. E_H is the concentration potential arising across membrane M_H and E_L is a similar potential arising across M_L. E is the potential across the complete electrochemical cell. E is the asymmetry potential developing across the whole cell when the activities a_1 and a_2 are equal.

cations), were considered to hold in between a high concentration of counterions associated with the polyelectrolyte (PSSA). In this arrangement, asymmetry potential arose as a result of the differences in the selectivities of the two membranes. Expressing selectivities in terms of counterion transport numbers, Lakshminarayanaiah and Siddiqi[5] showed that the emf E across the composite cell (Figure 3) was given by

$$E = \frac{RT}{F} \left[(2\bar{t}_{+(H)} - 1) \ln \frac{a_m}{a_1} - (2\bar{t}_{+(L)} - 1) \ln \frac{a_m}{a_2} \right] \quad (1)$$

where $\bar{t}_{+(H)}$ and $\bar{t}_{+(L)}$ are the transport numbers of counterion in high charge and low charge density membranes respectively, a_m is the activity of counterions existing between the membranes, and a_1 and a_2 are the activities of the two solutions contacting high and low charge density membranes. When $a_1 = a_2 = a$, the asymmetry potential is given by

$$E = \frac{2RT}{F} \left[\bar{t}_{+(H)} - \bar{t}_{+(L)} \right] \ln \frac{a_m}{a} \quad (2)$$

Lakshminarayanaiah and Siddiqi[5] showed that the potential E arising across the model system (Figure 3) conformed to Eqs. (1) and (2) and as a result they came to the conclusion that the magnitude of E (Eq. (2)) arising across composite membranes was controlled more by a_m, the activity of cations associated with PSSA and free electrolyte, than by the factor $\left[\bar{t}_{+(H)} - \bar{t}_{+(L)} \right]$. The results of

long term equilibration experiments and other results obtained with these membrane systems (see below) can be explained in terms of Eq. (2). Lack of good reproducibility of potentials observed with different pieces cut out of the same membrane can be attributed to the heteroporous structure of the membrane. In this type of structure, the different regions of the same membrane will trap different quantities of PSSA and electrolyte thereby giving different values for a_m. Similarly during long term equilibrations with dilute solutions, removal of trapped electrolyte will follow different pattern on the time scale depending on how big or small the membrane pores were. On the basis of this model, thorough washing and equilibration with dilute solutions of the composite membrane would give low values for the asymmetry potential. This is in agreement with our observations (see Figure 2). We always found short term equilibrations gave high potentials and long term equilibrations gave low values. Further use of membranes which were already used once before and therefore subject to washings and equilibrations always gave low values for the asymmetry potential. So long as the composite membrane retained a structure which stabilized the value for a_m, irrespective of the period of equilibration, steady and good potentials would be obtained. If the activity of trapped ions a_m is due to sorbed electrolyte which can be partially removed by washing and equilibration over a period of time, the potential would always decay with time as observed in the present studies. In view of this, it is not surprising to obtain with these membrane systems results which would not agree

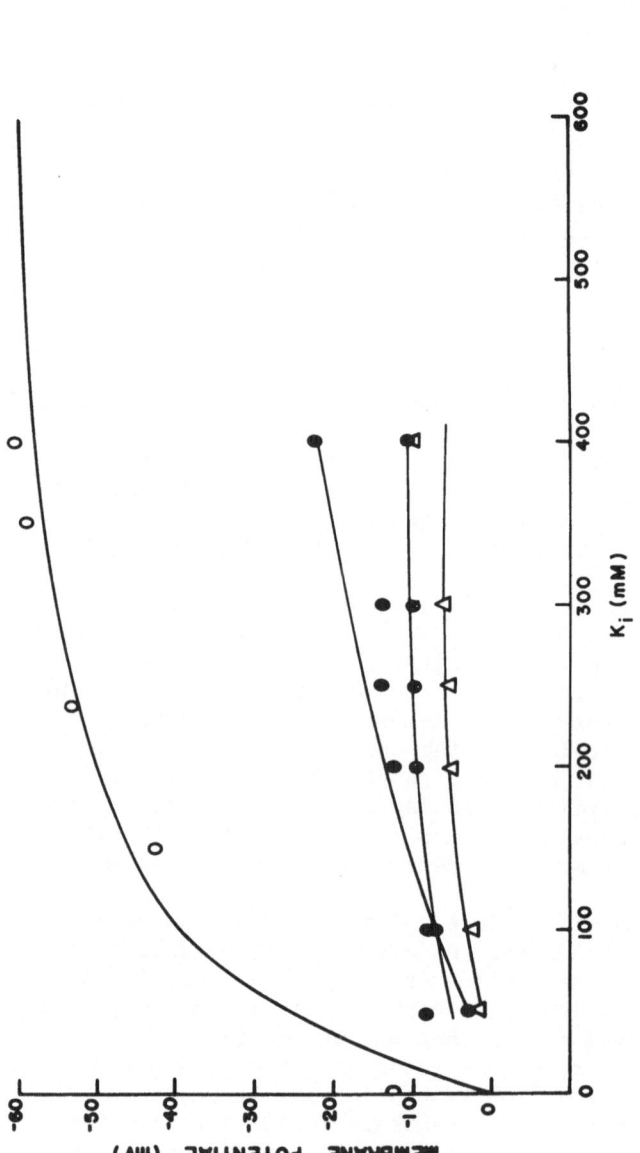

K_i (mM)

MEMBRANE POTENTIAL (mV)

Fig. 4. Membrane potential across different membrane systems plotted as a function of K^+ concentration. The high charge density side indicated to be negative. The solid line represents the data of Baker et al[7] who used the "perfusion technique" to replace the natural axoplasm of the giant axon of the squid with different isotonic solutions containing varying amounts of K_i^+. (O) represent the data obtained by Botre et al[3] with a composite membrane when the "outside" solution was 0.45 M NaCl and 0.01 M KCl and the "inside" solution contained different amounts of K_i. The ionic strength of this solution was kept at 0.45 M by replacing K^+ with Na^+. ($\bullet$), ($\circledcirc$) and ($\triangle$) represent data of this study using a genuine Liquori-Botre sandwich-type membrane, a layer-type membrane and the model system of Figure 3 when the middle compartment contained 0.5 M NaCl solution respectively. The solutions used were the same as those used by Botre et al.

with those obtained by others. For example in Figure 4 are given the results obtained by Botre et al[3] with a composite membrane. These potentials give good agreement with the resting potentials measured by Baker et al[7] using the perfusion technique on a squid nerve membrane. In these experiments the natural axoplasm of the giant axon of the squid was replaced by isotonic solutions containing varying amounts of K_i^+. In the same figure are given also some of our results obtained with three membrane systems, viz. the genuine Liquori-Botre sandwich-type membrane, our layer-type membrane and the model system (Figure 3) with 0.5 M NaCl solution in the middle chamber. The solutions used on either side of the composite membranes are indicated in the legend to Figure 4. Although our results with the three systems are consistent, they do not agree with the published values of Botre et al[3].

REFERENCES

1. A. M. Liquori and C. Botre, Ric. Sci., 34, (6), 71 (1964).
2. A. M. Liquori and C. Botre, J. Phys. Chem., 71, 3765 (1967).
3. C. Botre, S. Borghi and M. Marchetti, Biochim. Biophys. Acta, 135, 162 (1967).
4. A. M. Liquori, Il Farmaco (Pravia) Ed. Sci., 23, 999 (1968).
5. N. Lakshminarayanaiah and F. A. Siddiqi, communicated elsewhere.
6. R. Neihof, J. Phys. Chem., 58, 916 (1954).
7. P. F. Baker, A. L. Hodgkin and T. I. Shaw, Nature, 190, 885 (1968).

ACKNOWLEDGMENTS

The work has been supported by Public Health Service grant NB-08163-02.

INDEX

311